Engineering

OTHER TITLES OF INTEREST FROM ST. LUCIE PRESS

Contamination of Groundwaters

Environmental Fate and Effects of Pulp and Paper Mill Effluents

Lead in Soil: Recommended Guidelines

Naturally Occurring Radioactive Materials: Principles and Practices

Economic Theory for Environmentalists

Handbook of Trace Elements in the Environment

Emergency Response and Hazardous Chemical Management

Principles and Practices

Clyde B. Strong, M.S.
Clyde Strong and Associates
College Station, Texas

T. Rick Irvin, Ph.D.
Institute for Environmental Studies
Louisiana State University

Advances in Environmental Management Series

T. Rick Irvin
Executive Editor

Camilo Cruz-Batres
Associate Editor

Produced jointly by The Society of Environmental Management and St. Lucie Press

St. Lucie Press
Delray Beach, Florida

Printed and bound in the U.S.A. Printed on acid-free paper.
10 9 8 7 6 5 4 3 2 1

ISBN 1-884015-77-8

Direct all inquiries to St. Lucie Press, Inc., 100 E. Linton Blvd., Suite 403B, Delray Beach, Florida 33483.

Phone: (407) 274-9906
Fax: (407) 274-9927

StL

Published by
St. Lucie Press
100 E. Linton Blvd., Suite 403B
Delray Beach, FL 33483

PREFACE

Effective management of hazardous chemical use, storage, transport, and disposal is an increasing concern—and escalating cost—to chemical, petrochemical, and manufacturing firms. One effective tool in controlling costs associated with hazardous chemical usage is training both staff and management in the appropriate practices to be followed which minimize both the use of toxic chemical agents as well as the likelihood of uncontrolled releases of these agents into the environment. *Emergency Response and Hazardous Chemical Management: Principles and Practices* is intended to serve as a stand-alone review and course text for professionals managing the use of and responding to the release of hazardous chemical agents used by the industrial community.

This text is also designed to support the course and credential programs in environmental management and emergency response provided by the National Registry of Environmental Professionals (NREP). NREP is a not-for-profit professional organization providing registration which certifies demonstrated professional competence and ability in the environmental disciplines. NREP professional registration programs serve to document the qualifications that individuals are required to possess, as defined by environmental law, in order to carry out environmental management duties.

ACKNOWLEDGMENTS

We wish to thank our editor, Sandy Pearlman, for lending her tremendous expertise to working with us to bring this text to fruition. Her patience in working with two first-time authors was consistent and very much appreciated. Accolades go to Camilo Cruz-Batres, Associate Editor of the *Advances in Environmental Management Series,* for bringing his environmental expertise—and inexhaustible energy—to this project. Mark Schneider produced all of the graphics contained in this volume; his care and quality in customizing diagrams of varying content to the needs of this text have provided a particular resource to the reader. Special appreciation also goes to Dennis Buda, President of St. Lucie Press, for his diligent work with both authors to bring this, the first volume in the *Advances in Environmental Management Series,* jointly produced by St. Lucie Press and the Society of Environmental Management, to press. Lastly, we thank Mr. Dick Young, P.E., R.E.M., Executive Director of the National Registry of Environmental Professionals and Publisher, *Environmental Protection Magazine.* Dick has personally provided leadership and guidance to the development of many of the environmental credentialing and education organizations in the United States today. His leadership was central to bringing this text and text series, as well as the Society of Environmental Management, to reality.

AUTHORS

Clyde B. Strong, B.S., M.S., is President of Clyde B. Strong Associates. He has over 20 years of experience in environmental management, hazardous materials professional education, and emergency response program development. A member of numerous professional and honors societies, he received his B.S. degree from Texas Tech University and M.S. degree from Texas A&M University in 1977. The author of industry publications and guidebooks, Mr. Strong has designed and implemented continuing environmental and hazardous materials education programs for chemical and petrochemical manufacturing facilities, petroleum production sites, waste management facilities, universities, and national training centers in the United States, Canada, and overseas. Clyde Strong and Associates currently directs professional education programs in environmental and hazardous materials management for thousands of property and plant managers, technicians, scientists, and engineers at locations across the United States. Mr. Strong serves as Chairman of the Examination Committee for the Emergency Response Credential Programs of the National Registry of Environmental Professionals.

Dr. T. Rick Irvin, Ph.D., serves as Associate Professor at the Louisiana State University Institute for Environmental Studies and chairs the Academic Advisory Board of the National Registry of Environmental Professionals. Dr. Irvin received his B.S. degree *summa cum laude* from the University of Georgia and Ph.D. in toxicology from MIT in 1983. The author of academic, industry, and government publications, he has served as Principal Investigator on over two million dollars of research funded by government, industry, and private foundations in the areas of environmental toxicology, chemical carcinogenesis and developmental toxicology, environmental site assessment and risk assessment, and risk-based monitoring and remediation technologies for waste management. Dr. Irvin has developed undergraduate, graduate, and industry short courses on environmental site assessment, environmental science and toxicology, and waste management. He serves as a consultant to environmental, financial, and chemical firms on the management of toxic wastes and hazardous materials. Dr. Irvin also serves as Executive Editor of the text series *Advances in Environmental Science* and *Advances in Environmental Management* published by St. Lucie Press.

Camilo Cruz-Batres, M.S. R.E.P., is a Research Assistant at the Louisiana State University Institute for Environmental Studies. He received his B.S. degree in microbiology from LSU in 1991 and a master's degree in environmental studies from LSU in 1994. Mr. Cruz-Batres has conducted and directed validation of hazardous waste remediation technologies for complex mixtures of chemical wastes including wastes from the refining, oil production, and pulp and paper industries. He has also developed and validated laboratory toxicity test systems, employing *Ceriodaphnia* and *Photobacterium* sp., for identifying toxic constituents in hazardous waste mixtures. Mr. Cruz-Batres is currently serving as co-editor of *Microbial Toxicity Analysis of Environmental and Industrial Wastes.*

Introduction: Workplace Hazards During Chemical Emergency Response Actions

1.1 INTRODUCTION

Industrial activities all over the world generate large volumes of chemical wastes and by-products that must be transported, stored, and disposed of. Responsible parties in charge of handling these wastes must receive special training to ensure the safe management of the many dangerous situations created by the presence of hazardous substances. Management of hazardous wastes without proper preparation could result in grave consequences to human health and the environment.

A substance may be considered hazardous if it is flammable, explosive, toxic, corrosive, radioactive, cryogenic, or readily decomposes to give off oxygen at elevated temperatures. Although thousands of chemicals possess one or more of these properties, multiple hazards may not be apparent unless detailed information about a particular chemical is obtained. Chemical intermixing, furthermore, may create situations of magnified hazard if chemical interactions yield substances of higher flammability, toxicity, or any of the hazardous characteristics described above. Some of the dangers encountered at worksites include:

- Toxic substances
- Flammable materials
- Explosive materials
- Excessive noise
- Corrosive materials
- Biologically active materials such as bacteria and viruses
- Heat or cold stress
- Oxygen deficiencies
- Accidents resulting in physical harm
- Radioactive materials
- Cancer-causing agents

Workers in close proximity to these hazards may suffer varying degrees of health impacts depending on the type and degree of chemical exposure. Common consequences of toxic substance exposure include the following:

- Asphyxiation
- Poisoning
- Cancer
- Damage to liver, kidneys, nerve cells, etc.
- Harmful effects to the unborn child
- Loss of limbs
- Skin diseases
- Loss of hearing
- Eye injuries

1.2 HAZARDOUS CHEMICALS AND POTENTIAL EXPOSURE PATHWAYS

Chemical substances must enter an organism (plant or animal) in order to exert any toxic effects. Common routes of chemical absorption at the workplace include the following:

- **Inhalation**—breathing contaminated air
- **Dermal**—contact with harmful liquids, gases, solids, or contaminated clothing, equipment, medications, cosmetics, etc.

- **Ingestion**—eating or drinking contaminated food, water, or medications (food and cigarettes can be contaminated by gloves, equipment, or unwashed hands)

Exposure can result from any of the following situations:

- Lack of qualified personnel and/or the proper equipment to evaluate and define protection levels needed
- Improper selection of or insufficient training in the maintenance and use of personal protective equipment (such as respirators, special clothing, or safety glasses)
- Failure to follow instructions or wear prescribed protective equipment
- Failure or lack of engineering controls such as shields or drum handling equipment
- Unexpected hazards at the worksite
- Insufficient time to put on protective equipment in an emergency
- Walking unnecessarily through puddles or into vapor mists
- Failure to decontaminate immediately after splashes or spills occur

1.3 WORKPLACE PROTECTION

Workers in hazardous environments should familiarize themselves extensively with the kinds of chemicals and other potential dangers present at worksites. Common questions a supervisor can address include:

- What are the required protective clothing and equipment?
- What potential explosive and/or flammable conditions may be present?
- Is entrance into confined spaces necessary? (The air in these spaces must be checked for unsafe concentrations of airborne contaminants and for sufficient levels of oxygen.)
- What emergency equipment is available, where is it located, and how is it properly operated? (This equipment must be checked often and kept in good working condition.)
- What is the availability of standby personnel?
- What are the standard operating procedures (SOPs) for evacuation and rescue in case of an emergency?

- If conditions or situations are likely to change during the work period, how will workers be notified?
- What is the work/rest cycle for each task?
- What are the prescribed decontamination procedures?
- Will the buddy system be used?
- Are there any physical hazards present (i.e., high-voltage power lines, heavy equipment, unstable walls or shelves)?

Untrained personnel should avoid contaminated areas and potentially life-threatening materials or situations at all times.

1.3.1 Hazardous Waste Site Training Requirements

Special equipment, training, and precautions are required when managing hazardous wastes. Workers should stay alert, follow the SOPs and supervisor instructions, and use common sense at all times. Conditions to remain aware of at the worksite include:

- Any weather changes (hot or calm air may increase chemical concentrations and require additional protection)
- Wind direction (avoid dust and vapors by working upwind if possible)
- Odors that may indicate the presence of chemicals
- The location of someone who can help if an emergency arises
- SOPs for any necessary decontamination, including cleaning and storing or disposing of contaminated equipment and clothing (family and friends can be exposed to toxic substances carried on clothing, shoes, tools, etc.)
- Hand-washing requirements before eating, drinking, smoking, or using the restroom
- Contaminated clothes disposal and showering and changing needs prior to leaving the work area
- Areas designated for keeping food, drinks, cigarettes, and personal care items
- Heavy equipment operating near you
- The proper handling of drums and other equipment
- The need for proper personal protection equipment and its limitations
- Emergency procedures and the evacuation signal
- Where and how to exit from every area

1.3.2 Personal Protective Equipment

Personal protective equipment (PPE) requirements vary depending on the substances handled, existing conditions, and particular situations. Special suits, hard hats, goggles, face shields, aprons, boots, gloves, and respirators are examples of PPE designed for specific purposes. Proper operation of PPE requires an understanding of the inherent advantages and limitations and that all instructions and written procedures be followed.

Protective clothing may be used as a shield from fire, toxic chemicals, and/or corrosive materials. Such clothing includes splash suits, fully encapsulating suits, and chemical-resistant clothing. Whereas this type of PPE is designed to allow human access to hazardous environments by completely enclosing the body, other PPE protects only specific portions of the body. Fire-resistant clothing is not generally designed to provide a high level of protection against chemical exposure. Additional precautions, therefore, should be taken when exposure to both chemicals and fire is likely.

Insulation should be worn under protective clothing during winter months. Proper rest and cool-down periods, with body fluid and salt replacement, should be provided when protective clothing is worn during the summer or in hot conditions.

1.3.2.1 Splash Suits and Fully Encapsulating Suits

Splash suits and associated splash protection clothing are worn to keep hazardous materials from contaminating the body. Such clothing may include: gloves, boots, aprons, goggles, jackets, leggings, hoods, and coveralls made of chemically resistant materials composed of treated paper or specially formulated rubber. Different combinations of this type of protective clothing can provide the level of protection necessary for particular situations. Workers should know the location of emergency eyewash and shower facilities and familiarize themselves with decontamination protocols for reusable protective clothing. Workers should decontaminate themselves immediately after being splashed with hazardous chemicals.

Fully encapsulating suits are usually worn to protect the body against exposure to airborne concentrations of highly toxic or corrosive chemicals. (Atmosphere-supplying respirators must be worn with these suits.) Fully *encapsulating* does not necessarily mean fully *protective*, however, and because no suit is resistant to all chemicals, different suits are required for particular situations. Due to the specialized nature of this equipment, its use requires special training and experience, including plans for rescue and escape from the suits themselves.

1.3.2.2 Respirators

Contaminants absorbed through the lungs enter the bloodstream faster than through any other toxicant absorption route. Respirators prevent airborne contaminants from entering the body. Workers must know the proper use and limitations of such respirators; furthermore, effective respirator operation requires correct selection, fitting, and maintenance.

Exposure to high concentrations of a toxic substance, even for a short time, can cause serious injury or death. Low concentrations of certain toxic substances can cause permanent damage to the lungs, liver, kidneys, or other organs. Work environments, in addition, can be fatal if atmospheric oxygen concentrations are too low. Proper respirator selection must be based on hazard evaluation, therefore, including a determination of contaminant concentrations and forms and an assessment of atmospheric oxygen levels.

There are two basic categories of respirators: air-purifying respirators and atmosphere-supplying respirators. Air-purifying respirators are designed to remove specific air contaminates before they are inhaled. (WARNING: some contaminants cannot be removed by air-purifying respirators.) Respirators should not be used in situations that are immediately dangerous to life and health (IDLH) or when unknown contaminants are involved. Proper selection of an air-purifying respirator depends on:

- The contaminant to be removed from the air
- Contaminant concentration
- Respirator efficiency in removing the particular contaminant
- Contaminant warning properties

There are two types of air-purifying respirators:

- Filtering purifiers to remove dusts, mists, and fumes
- Sorbent purifiers to remove gases and vapors

Each type of respirator is accompanied by a variety of air-purifying filters, cartridges, and canisters designed to protect against specific chemical classes. Cartridges, for example, are small and usually attached directly to the facepiece. They remove low concentrations of pollutants most effectively; thus atmospheric contaminant concentrations must be known before determining whether cartridges are appropriate for the given situation. Using the wrong cartridge or canister or attaching different types of canisters on each side of the facepiece should be avoided at all times. Personnel employing powered air-purifying respirators should ensure batteries are charged.

Cartridges and canisters should be replaced at least once per day to prevent filter or sorbent material saturation; high contaminant concentrations may warrant more frequent changes. Air-purifying respirators should only be used against contaminants of known "warning properties" (odor, irritation, etc.). Failure to remove these warning properties is a clear sign of saturated cartridges; workers in this situation should immediately enter the clean area and replace cartridges or canisters according to specific supervisor instructions. Cartridges and canisters, furthermore, are color coded for their specific use. Typical cartridge color codes are as follows:

Atmospheric Contaminants	Assigned Color
Acid gas	White
Organic vapors	Black
Ammonia gas	Green
Acid gases and organic vapors	Yellow
Radioactive materials and highly toxic particles, except tritium and the inert gases	Purple
Pesticides	Chartreuse

Canister color for the contaminant is designated in the table above. Workers should always pay careful attention to the wording on the label.

Atmosphere-supplying respirators supply air to the facepiece from an uncontaminated (clean) air source. These respirators come in two basic types:

- Air-line respirators, which provide clean air to the face mask through a connecting hose from a large tank or tanks of compressed air or an air compressor located in a clean area.
- Self-contained breathing apparatus (SCBA), which provides clean air to the facepiece from an air cylinder carried on the worker's back. These respirators are used in the positive pressure mode during IDLH situations or in unknown atmospheres. Additional special training is required for their proper use and maintenance. Because this air supply is portable, it has only limited capacity, depending on particular breathing requirements. A warning signal is given when approximately 20–25% of the air remains. When the warning signal sounds, workers wearing the SCBA should leave the contaminated area immediately and proceed to a clean area to obtain a full cylinder; supervisors and buddies must be informed of the situation.

A qualified person should always be available at the worksite to identify safety and health hazards, establish the proper level of respiratory protection, and assist in respirator selection and fitting. Beards and certain facial hair are not compatible with respirators because they may affect the acceptable seating and sealing, allowing contaminated air to seep in. The temple pieces of regular eyeglasses can also interfere with the proper fit of full-facepiece respirators. For workers requiring corrective eyewear, special glasses can be fitted inside this type of facepiece; contact lenses, however, are in every case unacceptable.

Respirators may fail to work properly because of exposure to extreme heat or cold or after repeated use. Frequent maintenance checks should therefore be performed, and daily care and maintenance, including proper storage in a clean area, should be a regular part of the operation.

In general, wearing a respirator places an additional stress on the body. Preemployment physical examinations should determine whether a worker is physically capable of handling this additional stress; ongoing medical monitoring programs should be used to determine if continued respirator use is having an adverse effect.

1.4 MECHANICAL EQUIPMENT SAFETY

When handling containers of chemicals or removing chemically contaminated materials, heavy machinery operators may not be able to see all nearby personnel. Workers around heavy machinery must therefore be aware of objects and obstacles overhead at all times and keep the following rules in mind:

- Never walk under suspended loads.
- Never walk in front or in back of moving heavy equipment.
- Always be aware of heavy equipment location.
- Always wear a hard hat and proper foot protection.
- Do not operate any heavy equipment unless you are fully qualified and authorized to do so.
- Be aware that equipment can be a source of ignition for flammable or explosive materials.

Heavy equipment is not the only potential hazard when working with hazardous chemicals. Numerous smaller items such as pumps, compressors, generators, portable lights, drums, trucks, and hand tools are common additional dangers posed by hazardous waste sites. If not properly operated, these items can be as dangerous as larger equipment. Some rules to keep in mind in this case include:

- Be sure all machine guards are in place.
- Use equipment at its recommended speed and only for the jobs it was designed to perform.
- Always keep loose clothing away from moving parts.
- Never pump flammable material with gasoline- or electric-powered pumps. Use only hand- or air-powered diaphragm pumps and be sure to ground equipment and bond the containers.
- Use only nonsparking tools and be sure to ground equipment and containers when working in a flammable atmosphere or transferring flammable liquids.
- Be aware of the types of fittings on pumps and hoses. For example, acid and caustic will rapidly corrode aluminum.
- Check fluid levels (oil, fuel) periodically. Never add fuel to equipment while it is running.

1.5 HEAT STRESS

Hot environments can cause a variety of strains on the body, resulting in heat exhaustion or a potentially fatal heat stroke. Furthermore, PPE can significantly increase heat stress. Workers exposed to high-temperature conditions should learn to recognize symptoms of heat stress in themselves and co-workers. Employers should provide instructions on ways to reduce or prevent heat stress, such as frequent rest cycles to cool down and replace lost body fluids and salts. Symptoms of heat stress include the following:

- Clammy skin
- Weakness, fatigue
- Light-headedness
- Confusion
- Slurred speech
- Fainting
- Rapid pulse
- Nausea (vomiting)

If the above conditions of heat stress are evidenced, the following actions should be taken in the order given:

- Take the victim to a cooler and uncontaminated area.
- Remove protective clothing.
- Give drinking water, if conscious.
- Allow to rest.

Symptoms that indicate heat stroke include:

- Staggering gait
- Mental confusion
- Hot skin, temperature rise (yet may feel chilled)
- Convulsions
- Unconsciousness
- Incoherent, delirious

Heat stroke is a medical emergency, and if its conditions are evidenced, the following actions should be taken in the order given:

- Take victim to a cooler and uncontaminated area.
- Remove protective clothing.
- Cool the victim with water, cold compress, and/or rapid fanning.
- Transport the victim to a medical facility for further cooling and monitoring of body functions.

Additional information on heat-related problems can be found in most first aid books and in the DHHS Publ. No. 80-132, "Hot Environments," published by the National Institute of Occupational Safety and Health (NIOSH).

1.6 EMERGENCY PROCEDURES

SOPs should include plans for unexpected events such as accidents, fires, and explosions. Workers suspecting personal contamination should recognize the symptoms of overexposure to toxic substances and advise supervisors immediately. Symptoms may include:

- Irritation of skin, eyes, nose, throat, or respiratory tract
- Changes in complexion or skin discoloration
- Headaches
- Difficulty in breathing
- Nausea

- Dizziness or light-headedness
- Excessive salivation (drooling)
- Lack of coordination
- Blurred vision
- Cramps and/or diarrhea
- Changes in behavior

Personnel at a waste site should always know the location of emergency eyewash and shower facilities. Before entering confined spaces such as tanks or ditches (and periodically while remaining in such spaces), the air in the space should be tested by a qualified individual for oxygen content, explosive levels of gases, and contamination with toxic pollutants.

Workers wearing a respirator (SCBA) in an IDLH atmosphere must be accompanied by at least one more worker with a similar respiratory aid. Visual or verbal contact from a safe area must be maintained with those individuals at all times. Employers should devise a plan to ensure that everyone on site is protected from any likely incidents, and the necessary rescue equipment should be available in case of an emergency.

All personnel should be aware of site emergency rescue procedures and the locations of rescue equipment. Injured workers should be removed from an accident site as soon as possible to prevent further exposure. Help should be immediately solicited by rescuers in these situations, and emergency rescue procedures should be carefully followed.

1.6.1 Emergency Information

Telephones and two-way radios may provide a vital link between workers at the site of chemical release and should always be available at hazardous worksites, along with the necessary phone numbers for medical and other emergency services (local rescue squad, fire and police departments). Emergency first aid equipment and medical personnel, or someone who knows how to provide emergency first aid, should always be present and readily available at the worksite. Instructions for the fastest route to the nearest hospital or medical facility should also be available, along with necessary transportation equipment.

1.6.2 Contamination and Decontamination

When working with hazardous chemicals or chemical spills, it is important to establish and maintain "clean" areas at the site. Materials found in contaminated areas should be confined to specific "hot" zones whenever possible, and special decontamination zones and procedures should be used to help control the movement of hazardous materials from the hot zone.

Personnel should be familiar with all decontamination procedures at the worksite and should follow them carefully. After removing contaminated clothing, a worker should shower, wash his or her hair, and change into clean clothing in an uncontaminated area. Contaminated clothing or equipment should not be taken home because it can expose family and friends. Clothing, tools, and equipment should be decontaminated and stored or disposed of according to employer SOPs.

If spills or splashes occur in a contaminated area, the decontamination area should be accessed immediately to correct the problem. Clean protective clothing may be required in this case, before continuing work in the contaminated area.

1.7 MEDICAL SURVEILLANCE PROGRAM

Medical surveillance, an important part of an occupational health and safety program, is a way of keeping track of worker health through the use of periodic medical examinations and laboratory tests. The medical surveillance program can help a worker's doctor:

- Determine a baseline picture of worker health against which future changes can be measured
- Identify any underlying illness or conditions which might be aggravated by certain exposures to job activities
- Promptly recognize any abnormalities, toxic reactions, or other changes so that corrective measures can be taken

Medical records generated as part of the medical surveillance program are important aids doctors can access when diagnosing workers developing health problems that may be associated with exposure to hazardous substances. Abnormal conditions discovered early enough can allow for prescription of appropriate corrective actions that may prevent more serious conditions.

1.8 HEALTH AND SAFETY PROGRAM

The following ten points summarize elements of a sample program for worker health and safety:

1. Proper identification and qualification of the materials being handled
2. Constant surveillance of the work environment (e.g., a knowledge of weather conditions, contaminant levels, and fire/explosion potential)
3. Availability of protective and properly maintained equipment (both the PPE and engineering equipment to provide protection from and/or isolation of hazards)
4. An appropriate medical surveillance program, including a record of pre-employment conditions and work-related exposures
5. A fire and spill emergency control plan
6. A proper decontamination program (method for preventing unnecessary worker exposure and eliminating migration of contaminants from the site)
7. A comprehensive site work plan
8. A communication/safety program that keeps track of everyone on site and provides for medical, emergency, and/or community contacts
9. A site security plan for properly designating and controlling access to and exit from contaminated, decontaminated, and safe areas
10. A proper logistics plan (i.e., appropriate arrangements for eating, sleeping, washing and drinking water, compressed air, etc.)

Worker safety at a hazardous waste site can be maximized by keeping informed of the hazards involved, receiving the necessary training, following proper procedures and/or instructions, using required PPE, and remaining aware of conditions or situations at all times.

2

Chemistry and Properties of Hazardous Substances

2.1 INTRODUCTION

Uncontrolled hazardous chemical emergencies may place responding personnel at immediate risk because of the chemical, biological, or radiological properties of the materials released into the environment.

Each of these three groups of hazardous properties can be subdivided into a number of subcategories; for example, chemical hazards may be classified according to whether released agents include substances that are flammable, toxic, corrosive, or highly reactive. When evaluating the risk of a hazardous substance release, assessment personnel must keep in mind that one chemical can often exhibit more than one hazardous property. This section will review the hazardous properties of chemical agents often encountered during emergency response situations, including the physical and chemical characteristics of hazardous substances and the factors that modify or alter hazardous properties of chemical agents.

2.2 CHEMICAL REACTIONS

2.2.1 Introduction

Hazards associated with exposure to a chemical agent are generally governed by the chemical reactivity of the agent of interest. A chemical reaction, simply stated, is the conversion of one chemical substance to another. During simple combustion, for example, a substance such as methane (CH_4) is converted to carbon dioxide (CO_2) and water (H_2O) in the presence of oxygen.

In an emergency, there are a number of key factors a responder often considers to initially evaluate the hazards associated with a chemical substance. One key factor in evaluating the imminent hazard of an uncontrolled chemical release is whether a reaction is endothermic or exothermic:

- **Endothermic reactions** require heat for the reaction to be initiated or to continue.
- **Exothermic reactions** produce energy typically in the form of heat or light and are thus self-sustaining.

2.2.2 Chemical Reaction Rate

The rate at which a chemical reaction proceeds is important in evaluating how stable an emergency situation will be over time. A number of factors are initially important in determining the rate of chemical reactions during an emergency release:

- Surface area of the reactants
- Physical state of the reactants
- Concentration of the reactants
- Temperature
- Pressure
- Presence of a catalyst

An increase in contaminant surface area exposed (resulting when chemical agents spill from drums or other containers) generally leads to higher combustion rates because greater surface allows more contact with oxygen and provides a larger space where the reaction can take place. As an example, coal in its solid—or lump—form is combustible. However, if coal is finely ground as dust or powder, the resulting coal dust can become explosive in the presence of air.

The physical state of a released chemical substance (solid, liquid, or gas) also influences reaction rates as well as the stability of the substance in the environment. For example, most liquids and solids do not burn—it is the vapors produced from these materials that can burn. Flammable liquids ignite much more quickly when heated because the increased vapor concentration is a much more flammable material than the liquid state. The hazards from flammable liquids are thus often characterized according to the temperature—or flashpoint—below which these liquids will not produce sufficient airborne vapor levels to be ignited. Flammable gases, such as methane, can be ignited quite readily at normal temperatures provided sufficient oxygen is available. Several other important factors increase the likelihood that a released chemical substance will react in the environment:

- For a chemical reaction to take place, chemical molecules must come in contact with each other. When the concentration of a chemical reactant within a given space or volume is increased, the probability that two molecules will interact with each other is greatly increased. Therefore, as the concentration of a chemical agent in a specific volume increases, so does the rate at which a chemical agent will react with the environment.
- The application of heat increases chemical reaction rates because heat causes chemical molecules to become more energetic. In a more energetic state, chemical molecules move faster, increasing the probability that two molecules will contact each other and react. As a rule, for every 10°C increase in temperature of a chemical agent, the reaction rate increases by a factor of 2.
- A catalyst is a substance that speeds up a chemical reaction without being consumed in the reaction. For example, a platinum catalyst will cause a mixture of hydrogen and oxygen to explode; without the platinum catalyst present, this reaction would occur very slowly or not at all.

2.2.3 Chemical Compatibility

When materials can be mixed or stored together for extended periods of time without reaction, these materials are said to be compatible. Table 2.1 lists some

TABLE 2.1 Hazards Due to the Reaction of Incompatible Chemicals

Hazard	Incompatible Chemicals
Generation of heat	Sulfuric acid and water
Fire	Potassium permanganate and glycerin
Explosion	Metallic sodium and water
Toxic gas production	Nitric acid and copper
Flammable gas production	Calcium carbide and water
Formation of a substance with a greater toxicity than the reactants	Chlorine and ammonia
Formation of shock-sensitive compounds	Liquid oxygen and asphalt
Solubilization of toxic substances	Hydrochloric acid and chromium
Violent polymerization	Ammonia and acryonitrile

hazards due to the reaction of incompatible chemicals. The Environmental Protection Agency (EPA) has published a document, *A Method for Determining the Compatibility of Hazardous Wastes* (EPA 600/2-80-076), that may assist response personnel in determining compatibilities between chemicals.

2.2.4 Chemical Properties

Chemicals possess inherent properties important to determine the potential dangers of a substance released into the environment. Table 2.2 provides a comparison of important physical characteristics for some flammable liquids, and the following definitions discuss some chemical characteristics important in emergency response management.

- **Solubility**—The ability of a solid, liquid, or gas to dissolve in a solvent. Water solubility should be among the first considerations by responders at the scene of an emergency release, for a number of reasons. If a material is water soluble, it may be possible to dilute the chemical to the point where it no longer presents a hazard. Consideration must be given to:
  - The volume of water necessary to produce this dilution effect
  - The potential environmental problems that might result
  - Disposal of larger volumes of hazardous waste
- **Specific gravity**—The ratio of the density of a substance (at a given temperature) to the density of water (at 4°C). The density of water is 1.0 g/cc, but specific gravity is stated as a pure number with water = 1.0. Substances with a specific gravity greater than 1.0 will sink in water; substances with a

TABLE 2.2 Properties of Some Common Flammable Liquids

	Flashpoint	Flammable Limits	Specific Gravity	Vapor Density	Boiling Point	Water Soluble
Acetone	0°F	2.6–12.8%	0.8	2.0	134°F	Yes
Amyl acetate	77°F	1.0–7.5%	0.9	4.5	290°F	Slightly
Carbon disulfide	–22°F	1.3–44.0%	1.3	2.6	114°F	No
Ethyl alcohol	55°F	4.3–19.0%	0.8	1.6	173°F	Yes
Ethyl ether	–49°F	1.9–48.0%	0.7	2.6	95°F	Slightly
Gasoline	–45°F	1.4–7.6%	0.8	4.0	100–400°F	No

specific gravity less than 1.0 will float. If a substance floats, addition of water can cause a material to float and spread over a larger area. On the other hand, addition of water can smother a fire when the burning substance is heavier than water. Heavier-than-water materials that spill into a watercourse can be very difficult to observe, contain, and recover.

- **Vapor density**—A comparison of the density of a gas or vapor to the density of ambient air standardized to a value of 1.0. If the density of a vapor or gas is greater than 1.0, the substance will tend to settle and not disperse as readily as a gas or vapor with a density less than 1.0. Heavy gases or vapors may concentrate in low areas and dilute the oxygen concentration. Heavy gases can thus create dangerous explosion risks and toxic hazards because of their ability to linger in an area.
- **Vapor pressure**—The pressure a vapor exerts on the sides of a closed container. Vapor pressure is traditionally expressed in millimeters of mercury at a specific temperature. Vapor pressure increases with vapor temperature, which means that more liquid evaporates or vaporizes with increased temperature. Liquids with low boiling points have greater vapor pressure at a given temperature; the greater the rate of evaporation, the greater the potential hazard.
- **Boiling point**—The temperature at which the vapor pressure of liquid equals atmospheric pressure. The lower the boiling point, the more volatile a chemical substance. In evaluating the risk from fire, a lower boiling point generally means a greater danger. One other important point derived from the boiling point of a hazardous liquid is the likelihood of intake. Toxic liquids with a high boiling point tend to enter the body mostly through skin contact. With low boiling point liquids, the most common and serious route of exposure is inhalation.
- **Flashpoint**—The minimum temperature at which a substance produces enough vapors to form an ignitable mixture with the air. Relative flammabilities of various substances can be based on their flashpoint:
  - Flashpoint of 100°F or less: highly flammable
  - Flashpoint greater than 100°F but less than 200°F: moderately flammable
  - Flashpoint greater than 200°F: relatively inflammable
- **Flammable limits**—The concentration range of a combustible vapor/air mixture below which the mixture has too little fuel to ignite and above which there is too much fuel and too little oxygen to ignite. It is only within the flammable limits that an ignitable concentration of fuel and air exists. Typically an upper and a lower flammable limit will be given in the data sheets on a hazardous chemical. Table 2.2 illustrates this principle. In

general, the wider the range between the upper and lower flammable limits, the greater the risk of fire or explosion.

2.3 BLEVE (THERMAL RUPTURE)

BLEVE is an acronym for boiling liquid expanding vapor explosion. This term has been widely used in emergency response circles to describe the heat-induced rupture of a container of liquefied gas; a more correct term is "thermal rupture."

Fire contacting a tank below the level of the liquid has its heat absorbed away from the tank shell by the liquid. The liquid absorbing the heat may boil and build up pressure, but the integrity of the tank metal is not damaged. If the fire contacts the tank above the liquid level, the vapors in the tank do not absorb sufficient heat to produce cooling of the tank shell. The tank metal with sufficient application of heat becomes "plastic" and begins to thin or stretch due to the internal tank pressure. A bulge may appear and a tear develops; the tear opens the tank, causing a violent rupture. The power of the rupture is directly related to the amount of chemical remaining in the container.

When liquid in the container absorbs heat, boiling can occur, leading to vaporization and pressure build-up in the tank. With sufficient pressure, the relief valve may activate, releasing the tank pressure and contained chemical. Each time this occurs, the liquid level in the tank drops and the liquid cools. A substantial drop in the level of the liquid could cause a previously "wetted" area of the container to become dry and unable to transfer heat away from the metal. Stated simply, over time, as the relief valve operates, the liquid level will drop, exposing more of the tank metal to the full impact of the heat.

The application of water can cool the surface of a container, removing heat from the metal and maintaining metal strength. Extreme caution should be used in employing this technique, however. If flame impingement in the vapor space occurs for a sufficient period of time, multiple applications of water may be needed over extended periods before the situation stabilizes. Under these conditions, consideration should be given to the use of unmanned monitors.

2.4 HAZARDOUS CHEMICAL PROPERTIES

Because there are many thousands of hazardous chemicals in use by the chemical industry, only the more common hazards presented by various categories or classes of chemicals will be discussed. An attempt will be made here to provide insights into the multiple hazards exhibited by various common chemical groups.

2.4.1 Organic Chemicals

Compounds that contain carbon structure are generally referred to as organic compounds. Organic compounds are the most common chemicals found in living matter—plant and animal. Not surprisingly, by-products of animal and plant matter, such as fossil fuels, are also composed principally of organic compounds.

Most organic compounds are flammable—under room conditions in the presence of heat these compounds will burn. Compared with metals (also known as inorganic compounds), organics melt and boil at relatively low temperatures. Many organic compounds volatilize at room temperatures and have low ignition temperatures. Because the heat of combustion for these compounds is high, combustion radiates excess heat, causing the risk of igniting adjacent combustible matter. Organic compounds are relatively insoluble in water as well; thus, applying water to a fire supplied with organic fuel will cause the burning material to float over a large area. Fire-fighting foams and other extinguishing agents are usually more effective on fires involving most organic compounds.

Because organic chemicals are easily combustible, these agents should not be allowed to come into contact with oxidizing agents. Contact of organics with oxidizing agents will usually result in spontaneous ignition and intense fires due to the flammable nature of the organic and the supply of oxygen from the oxidizer. As discussed in the subsequent sections on toxicity, organic compounds are among the most hazardous toxic agents to man and the environment.

2.4.2 Oxidizing Agents

Oxidizing agents are chemicals that tend to seek or *grab* electrons from other elements or compounds. This reactive nature makes them particularly hazardous when they come into contact with organic compounds and many other chemicals. When reactions involving oxidizers occur, considerable heat is generated and absorbed by surrounding materials. In many cases, this heat is sufficient to cause ignition of combustible materials.

Oxidizing agents that contain oxygen in their molecular make-up are inherently unstable when exposed to heat and decompose to yield oxygen. This release of oxygen can greatly intensify a fire, yet with most oxidizers, addition of flooding amounts of water generally cools the material and dilutes the oxidizer.

Some oxidizers such as calcium hypochlorite should not be mixed with hydrochloric (muriatic) or other acids, because a product of the reaction is chlorine gas. Inorganic peroxides including hydrogen peroxide exhibit similar reactive characteristics. Hydrogen peroxide is very unstable, decomposing readily to yield oxygen. Concentrations over 30% require addition of a stabilizer because contact with organics under these conditions always leads to fire. Hydrogen peroxide, therefore, is a serious corrosive health hazard (human bodies are composed of many organic materials). Water is an effective extinguishing agent; nevertheless, but it should be used with caution because it causes inorganic peroxides to give off oxygen. Flooding amounts should be used to extinguish an inorganic peroxide fire, and the oxygen should be allowed to dissipate.

2.4.3 Water-Reactive and Pyrophoric Chemicals

Pyrophoric substances are those that ignite spontaneously in air without an ignition source. Ignition could result from a number of factors such as reaction with oxygen or water vapor in the air (water-reactive pyrophorics) or exposure to ambient temperature levels. Materials that ignite easily when rubbed or scraped are also considered pyrophoric.

Water can react with chemical substances in many ways, some of which are potentially hazardous. Chemicals that remove moisture from the air (hygroscopic substances), for example, should not be left in open containers because they can actually absorb enough moisture from humid air to overflow (sulfuric acid is a good example).

A more serious hazard generally occurs when water and a reactive chemical undergo a reaction called hydrolysis. The hazardous features of this reaction vary with the chemicals involved, but it can lead to the generation of flammable products, toxic gases, or corrosive materials.

The alkali metals lithium, sodium, and potassium react with water to form hydrogen gas and lithium hydroxide, sodium hydroxide, and potassium hydroxide, respectively. In the case of sodium and potassium, the heat of reaction and generation of hydrogen may cause ignition and possibly an explosion. Hydroxide solutions of potassium and sodium are highly caustic as well.

Other water-reactive groups include the organo-metallic compounds, sometimes grouped as alkylaluminums; some hydrides, such as boranes and silanes; some carbides; peroxides; some inorganic chlorides; and some organic compounds, such as acetic anhydride and acetyl chloride.

White (yellow) phosphorus is not water reactive and in fact is shipped and stored under a blanket of water. It is, however, considered pyrophoric. The ignition temperature is only 86°F, and even when ambient temperatures are low, rubbing or friction can cause ignition. White phosphorus burns producing tremendous volumes of white smoke, which, although it is slightly irritating, is not considered toxic.

2.4.4 Corrosives

Corrosive substances are one of the most commonly encountered groups of hazardous materials. The destructive action of these chemicals on living tissue is their primary hazard; therefore, personnel should have a clear understanding of the dangers associated with working near corrosive chemicals.

Corrosive substances can be categorized as acids or bases. Acids are compounds that yield hydrogen ions (H^+) when dissolved in water, and they have a pH of less than 7.0. They can evidence almost every hazardous characteristic exhibited by other chemicals. Acids can react with many metals to produce hydrogen gas, and in a confined space, such as a container, this could lead to an explosive atmosphere or a container rupture.

Picric acid is highly explosive and acrylic acid can undergo violent polymerization. Some acids, such as nitric and sulfuric acid, can be strong oxidizing agents, increasing the risk of ignition. Sulfuric acid (water reactive) liberates tremendous amounts of heat when diluted with water, and this mixing can lead to localized boiling and violent spattering. Chlorosulfonic acid, furthermore, releases heat and dense irritating fumes upon contact with water.

Some acids, such as carbolic acid (phenol), undergo reactions with other materials to produce highly toxic reaction products. Sulfuric acid reacts with various cyanide salts to liberate hydrogen cyanide gas. Nitric acid reacts with metals such as copper to give off toxic nitrogen dioxide vapors. Acids can also be flammable and initiate combustion—organic acids contain carbon and can be ignited. Inorganic acids do not burn, but the heat of reaction produced upon contact with certain materials is sufficient to cause ignition.

The terms alkaline, caustic, and base are generally used interchangeably to describe a group of compounds that supplies hydroxide (OH^-) ions when dissolved in water. Bases are on the opposite end of the pH scale from acids, having a pH greater than 7.0. Strong bases are usually considered to have a pH of 12.5 or greater and also exhibit a number of hazards, including corrosivity.

Bases react with acids (and vice versa) in a neutralization reaction, forming a salt and water. This reaction is accompanied by the evolution of considerable heat and may also produce spattering and frothing. Bases react with many metals to produce hydrogen, but typically the reaction is too slow to build up explosive concentrations of hydrogen. Sodium hydroxide and potassium hydroxide undergo exothermic reactions with water, liberating considerable heat.

2.4.5 Biological Hazards

Biological hazards refer to those agents that can cause disease or infections in people exposed. In transportation, they are often referred to as etiologic agents. Although emergencies involving these agents are rare, the transported containers of various hazardous wastes from laboratories and hospitals can potentially expose individuals to such hazards.

Biological agents, like many hazardous chemicals, can be dispersed through the air and water and may also be infectious through skin contact (typically broken skin). Personal protective equipment (PPE) can prevent exposure in many instances; in addition, showering, washing of the hands and face, and good personal hygiene are good protective measures.

Examples of biological hazards often overlooked include rabid animals, poisonous plants or animals, and disease-bearing hosts such as ticks carrying Rocky Mountain spotted fever.

2.4.6 Radiation Hazards

Unlike many other hazardous chemicals which have properties that can alert personnel to possible hazards, radioactive materials possess no warning signals. Alpha, beta, and gamma radiation is absorbed by the body unknowingly, and the primary action of radiation is to ionize materials through which it passes.[1] The ionization of molecule "A" can be represented as follows: $A \rightarrow A^+ + e^-$ where $\rightarrow$ indicates that the reaction is initiated by high-energy radiation. The ion, A^+, in the vicinity of electrons (e^-) may acquire one and become energized or excited, illustrated as $A^+ + e^- \rightarrow A^*$. Now that A^+ has become A^* with much more energy, it may form two totally new molecules. Thus, $A^+ + e^- \rightarrow A^* \rightarrow B + C$. This may not appear to be a significant hazard until you consider that this reaction is occurring in large molecules in the cells of the body. This mechanism can drastically alter the chemical activity of the body at the cellular level.

There are three types of harmful radiation of concern to response personnel. Alpha particles are positively charged, consisting of two protons and two neutrons. They have mass and energy but are easily absorbed and travel only 1–3 in. in air. A sheet of paper is sufficient to absorb alpha particles. The primary hazard they pose is from inhalation or ingestion.

Beta particles are electrons or other small particles called positrons. They are smaller than alpha particles and move at higher speeds. They can travel 10–100 ft in air. Few beta particles can penetrate beyond $^1/_2$ in. into the body, although in high concentrations, skin burns can result. The primary routes of absorption are through inhalation and ingestion.

Gamma radiation is pure electromagnetic energy and is wave-like rather than particulate. Gamma waves are an extreme external hazard because they travel great distances and can penetrate anything to a certain extent. Protective equipment is not sufficient to prevent harm from this type of radiation. PPE and good hygiene can be effective in reducing the potential effects of alpha or beta radiation, but keeping a safe distance is the only way to protect personnel from a source of gamma radiation.

2.5 REFERENCES

1. E. Meyer. *Chemistry of Hazardous Materials*. Prentice Hall, Englewood Cliffs, NJ, 1977.
2. U.S. EPA. *Hazardous Materials Incident Response Training Program*, 1983.
3. J. H. Meidl. *Flammable Hazardous Materials*. Glencoe, Encino, CA, 1978.
4. Association of American Railroads. *Emergency Action Guides*. Washington, DC, 1984.
5. L. Shieler and D. Pauze. *Hazardous Materials*. Van Nostrand Reinhold, New York, 1976.

3

Toxicology Principles

3.1 INTRODUCTION

A toxic substance is one that can destroy the life or harm the health of a living animal or plant; toxicology, in turn, is the scientific study of poisons and all chemical agents that produce these effects. Every substance has the ability to act as either a poison or a remedy, including water. The dose of chemical absorbed by an organism, however, and the length of exposure are what generally determine whether a given exposure will result in harmful or beneficial effects. This chapter will describe some of the mechanisms that lead to toxicity and how they relate to safety at the workplace.

3.2 GENERAL TERMINOLOGY

Toxic substances and their relative hazards can be ranked according to information obtained from animal tests. These tests are conducted by exposing sets of mice or other rodents to different doses of a chemical, monitoring the toxic effect elicited by each dose, and plotting the data on a graph to yield a dose–response curve. Although rodent toxicity data cannot be directly extrapolated to determine the effects a chemical exposure will have on humans, they do enable toxic substances to be rated according to the relative risks they pose. Relative toxicities thus are listed according to the concentration (given in milligrams of toxin per killigrams of body weight, or ppm) expected to provoke a certain reaction in an "average" person. Individual responses to the same concentrations, however, can vary.

Dose–response curves and relative toxicities are some of the general terms employed when describing toxins. Before discussing the different modes of toxic action, some other important terms need to be defined.

- **Threshold limit value–time-weighted average (TLV-TWA)**—The per-hour average concentration of a substance for an 8-hr work day per 40-hr work week to which nearly all workers may be repeatedly exposed without adverse effects.
- **Threshold limit value–short-term exposure limit (TLV-STEL)**—A 15-min time-weighted exposure that should not be exceeded at any time even if the 8-hr average chemical concentration is within the TLV. Exposures at the STEL should not be longer than 15 min or more often than 4 times per day. Sixty minutes should be allowed between successive STEL exposures.
- **Threshold limit value–ceiling (TLV-C)**—The concentration that should not be exceeded at any time.
- **Toxic dose low (TD_{Lo})**—The lowest dose of a substance introduced by a route other than inhalation and reported to produce a toxic effect in humans or carcinogenic, neoplastigenic, or teratogenic effects in laboratory animals.
- **Toxic concentration low (TC_{Lo})**—The lowest concentration of a substance in air reported to produce a toxic effect in humans or carcinogenic, neoplastigenic, or teratogenic effects in laboratory animals.
- **Lethal dose low (LD_{Lo})**—The lowest dose, other than the LD_{50}, of substance introduced by any route other than inhalation that has been reported to cause death in humans or laboratory animals.
- **Lethal dose fifty (LD_{50})**—An estimated dose of a substance that kills 50% of a defined experimental animal population.
- **Lethal concentration fifty (LC_{50})**—An estimated concentration of a substance in air expected to cause death in 50% of a defined experimental animal population, after a defined exposure time.
- **Lethal concentration low (LC_{Lo})**—The lowest concentration of a substance in air, other than the LC_{50}, that is known to cause death in humans or laboratory animals.
- **Immediately dangerous to life and health (IDLH)**—The maximum concentration from which one could escape within 30 min without irreversible health effects.

3.3 FACTORS INFLUENCING TOXICITY

Many factors influence the toxic endpoints of a given chemical exposure, and these should be considered when assessing the workplace hazard of different chemical substances. Depending on the route of exposure, for example, the toxic hazard of a substance such as snake venom may range from insignificant (when ingested) to deadly (when injected underneath the skin). Some of the principal factors influencing toxicity will be discussed in the following sections.

3.3.1 Route of Chemical Absorption

In order for a chemical substance to produce a toxic effect, it must be absorbed by an organism. The route of exposure, therefore, is an important factor influencing the spectrum of toxic endpoints a chemical may elicit. There are four routes through which a toxic substance may enter the body:

- Inhalation
- Dermal contact
- Ingestion
- Subdermal injection

Inhalation is the most potentially serious route of exposure because the respiratory system is constantly exposed to the environment and is therefore difficult to protect. The lungs have an exposed surface area of between 70 and 100 m^2, compared with 2 m^2 of skin surface and 10 m^2 of exposed area in the digestive system. The amount of chemical that may enter the body through inhalation, therefore, is 10–50 times higher than through other routes. In addition, alveoli of the lungs are continually exchanging gases with the bloodstream, so that a toxicant absorbed through the lungs is almost immediately deposited throughout the body.

The skin functions as a barrier that prevents foreign substances from entering the body. Hence, chemical absorption through this route does not occur readily. Absorption may be enhanced when the skin is broken, hydrated (softened due to prolonged contact with water), or hot (sweat dissolves some solids). Toxic action may take place directly on the skin or in other organs after a toxin has entered the bloodstream.

Exposure through ingestion is not as common as the first two routes, but it can occur by transmittal to the mouth from hands, clothing, or other objects. Particulate toxins, furthermore, are trapped in the throat by mucus and then swallowed. Once in the stomach, toxins may be chemically modified by gastric enzymes and acids and subsequently absorbed through the intestinal walls and deposited in the bloodstream.

Subdermal injection is the least common route of chemical absorption at the workplace, and it may take place by accident. Contact with sharp objects at a waste site could potentially expose workers. In most cases, common safety precautions should prevent this mode of chemical absorption.

3.3.2 Toxicant Physical State and Synergisms

Industrial chemical substances in the workplace are often found in several physical states, in combination with other chemicals, or as complex chemical mixtures. The toxic hazard of a chemical substance varies significantly depending on its physical state—solid, liquid, or gas. In general, gaseous substances present the greatest toxic danger because they can enter the human body through the lungs. Solids, conversely, are not as easily absorbed; unless they change their physical form, potential routes of solid toxicant absorption are limited to ingestion.

Complex chemical mixtures are usually a greater toxic hazard than any pure substance alone. Interactions between single components in a chemical mixture may lead to an amplification of the original individual toxicities. Synergistic reactions, the result of chemical mixing, can be categorized as:

- **Multiplicative**—The toxicities of the original components are greatly enhanced when intermixed.
- **Additive**—Overall mixture toxicity is the sum of the individual component toxicities.
- **Antagonistic**—The toxicity of individual mixture components is greatly reduced in the presence of other chemicals.

3.3.3 Sex, Age, Genetic Make-Up, and Health Condition

Some chemical substances may exhibit selective or enhanced toxicity according to the sex of the exposed individual. Women, in general, tend to have more body fat and are thereby able to accumulate higher concentrations of fat-soluble substances. Chemicals stored in fat deposits may slowly re-enter the bloodstream over time and lead to long-term toxic effects.

Chemical toxicity may also be age selective. A person's blood chemistry, metabolism, and excretory patterns, for example, change over time, effectively modifying one's ability to deal with a toxic exposure. Children, in addition, tend to have higher respiratory rates than adults and therefore have different responses to certain types of chemicals.

Genetic make-up is another factor that affects individual reactions to toxic substances. For example, some people may inherit systemic weaknesses or lack the ability to produce certain enzymes, thereby making them more susceptible to the effects of toxins.

Persons in good physical condition, furthermore, tend to have more strength and stamina and have lower heart and respiratory rates under the same work load. All these factors need to be considered when assessing workplace safety. Table 3.1 provides a classification of factors influencing toxicity.[1]

TABLE 3.1 Factors Influencing Toxicity

Type	Examples
Factors related to the chemical	Composition (salt, base, acid), physical characteristics (particle size, liquid, gas, solid), physical properties (volatility, solubility, ignitability), presence of impurities, breakdown products
Factors related to exposure	Dose, concentration, route of exposure, exposure duration
Factors related to the person exposed	Heredity, immunology, nutrition, hormones, age, sex, diseases
Factors related to the environment	Carrier (air, water, food, soil), chemicals present (synergism, antagonism), temperature, air pressure

Source: U.S. EPA. *Hazardous Materials Incident Response Training Program*, 1981.

3.4 MAJOR CLASSES OF TOXICANTS ACCORDING TO PHYSIOLOGICAL REACTIONS

The physiological effects of various types of toxic substances and how different organ systems in the body respond are examined here. Three major classes of toxicants can be derived from the physiological action of toxins:

- Irritants
- Asphyxiants
- Systemic poisons

3.4.1 Irritants

Irritants are chemicals that inflame the mucous membranes and air passages of the respiratory system, causing airways to constrict, which makes breathing difficult. Irritant action is a function of the affinity of a material for water, and as a general rule, the concentration of the irritant is more critical than the duration of exposure. Irritants, therefore, may be placed in one of three categories depending on their water solubility characteristics:

- Upper respiratory irritants
- Lower respiratory irritants
- Whole respiratory system irritants

Upper respiratory irritants are quite water soluble; consequently, their effects are primarily evidenced in the nose, throat, and bronchi. Ammonia, a common irritant, produces extreme burning sensations at less than lethal concentrations and thereby exhibits good warning properties that allow workers in a contaminated area to evacuate promptly. Lower respiratory irritants are less water soluble than upper respiratory irritants and, in many cases, do not exhibit adequate warning properties before dangerous concentrations are encountered. The primary effects of lower respiratory irritants occur in the alveoli and other regions of the lungs. Nitrogen dioxide, a classic example of a lower respiratory irritant, slowly mixes with moisture in the lungs to form nitric acid. There have been reported cases in which few noticeable symptoms appeared until 5–24 hr after exposure to the gas. Whole respiratory system irritants are moderately water soluble and cause inflammation throughout the entire respiratory system. Warning properties for these substances are usually adequate, so that lethal concentrations are rarely accidentally encountered.

Figures 3.1 to 3.3 illustrate the areas of the respiratory tract and the classes of irritants that may affect them.

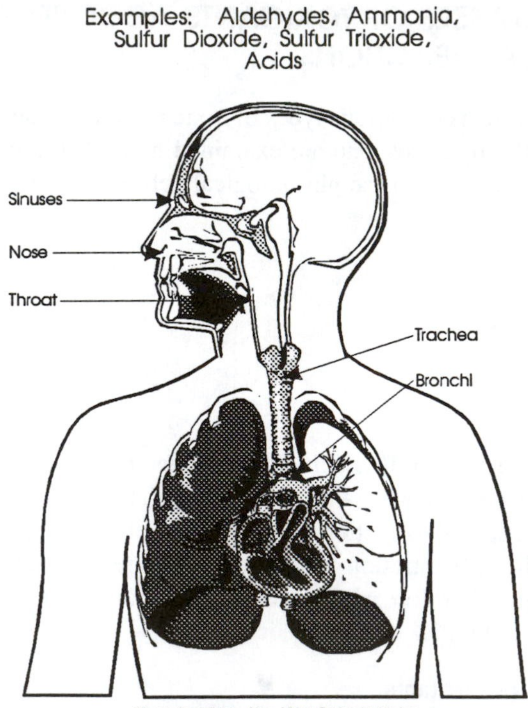

FIGURE 3.1 Upper Respiratory Irritants

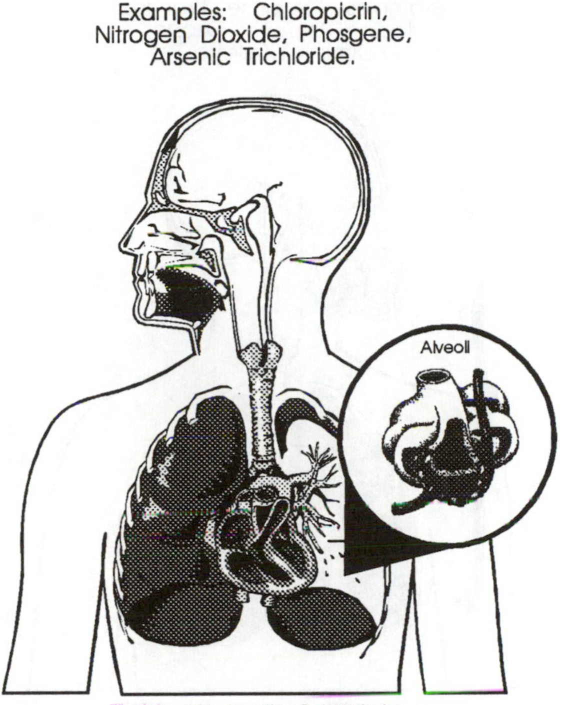

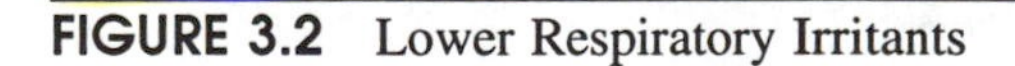

FIGURE 3.2 Lower Respiratory Irritants

3.4.2 Asphyxiants

Asphyxiants are substances that deprive body tissues of oxygen. Two categories of asphyxiants exist:

- Simple asphyxiants
- Chemical asphyxiants

Simple asphyxiants are physiologically inert and produce no direct harmful effects; they simply displace oxygen carried by red blood cells. Examples include nitrogen, helium, neon, and hydrogen. These gases are often deadly because they do not have an odor and can therefore overcome a person without their being aware of it. Response personnel and site workers should be able to recognize the following symptoms of acute lack of oxygen: mild euphoria (light-headedness), increased heart and respiratory rates, headache, fatigue, altered breathing, and nausea. Table 3.2 describes the physiological effects of various levels of oxygen deficiency.

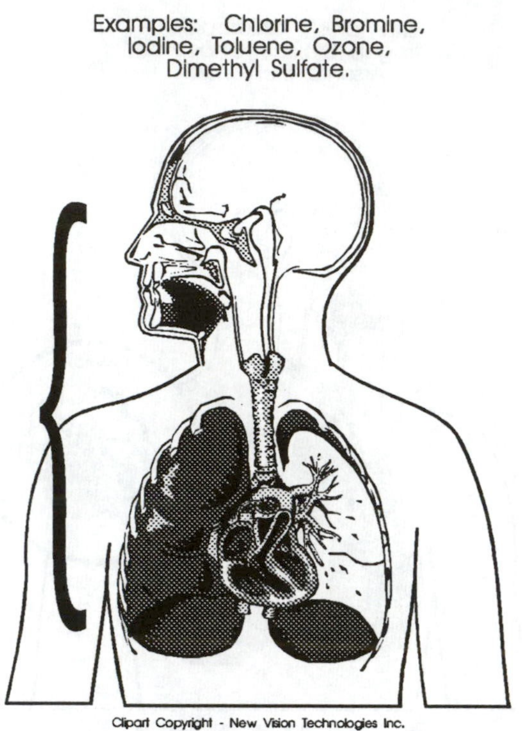

FIGURE 3.3 Whole Respiratory Irritants

TABLE 3.2 Physiological Effects of Oxygen Deficiency

% O_2 at Sea Level	Effects
21	Nothing abnormal
16–21	Loss of peripheral vision, increased breathing volume, accelerated heartbeat, impaired attention and thinking, impaired coordination
12–10	Very faulty judgment, very poor muscle coordination, muscular exertion causes fatigue that may cause permanent heart damage, intermittent respiration
10–6	Nausea, vomiting, inability to perform vigorous movement or loss of all movement, unconsciousness followed by death
<6	Spasmodic breathing, convulsive movement, death in minutes

TABLE 3.3 Toxic Effects of Carbon Monoxide

CO Concentration (ppm)	Effects
35	No effects for average working lifetime
200	Mild headache after 2–3 hr of exposure
400	Headache and nausea after 1–2 hr of exposure
800	Headache, nausea, and dizziness after 45 min of exposure; collapse and possible unconsciousness after 2 hr of exposure
1,000	Unconsciousness after 1 hr of exposure
1,600	Headache, nausea, and dizziness after 20 min of exposure
3,200	Headache, nausea, and dizziness after 5–10 min of exposure; unconsciousness after 30 min of exposure
12,800	Immediate effects; unconsciousness, danger of death after 1–3 min of exposure

Chemical asphyxiants prevent body tissues from getting enough oxygen despite the presence of adequate atmospheric oxygen levels. Some chemical asphyxiants include carbon monoxide, cyanide compounds, hydrogen sulfide, and aniline. Carbon monoxide binds to blood hemoglobin over 200 times more readily than oxygen. Hemoglobin affinity for carbon monoxide can effectively prevent an adequate supply of oxygen from being transported to body tissues. Table 3.3 lists some concentrations and symptoms for carbon monoxide exposure.

Aniline and other aromatic amines, nitrates, aromatic nitro-compounds, dichromates, and perchlorates modify the oxygen–hemoglobin bond in a way that prevents oxygen from being released into tissues. A symptom of this type of toxic exposure is a bluish discoloration of the lips and fingernails. Hydrogen sulfide, through a mechanism not completely understood, acts on the central nervous system to stop respiration. Hydrogen cyanide acts at the cellular level through enzyme inhibition to prevent cells from receiving and using oxygen from the blood.

3.4.3 Systemic Poisons

Specific organs and organ systems, including the central nervous system, liver, kidneys, and heart, can be adversely affected by exposure to toxins. Poisons that affect the central nervous system may be divided into two groups:

- Depressants
- Convulsants

Depressants, characterized by many types of hydrocarbons, produce narcotic-like symptoms by depressing the central nervous system, possibly leading to unconsciousness and coma. An individual who becomes intoxicated with ethyl alcohol, for example, will exhibit immediate symptoms of central nervous system depression. Table 3.4 lists some central nervous system depressants.

TABLE 3.4 Central Nervous System Depressants

Lead compounds	Ethyl ether
Isopropyl alcohol	Ethylene
Ethyl acetate	Ethylene dichloride
Cyclohexane	Methyl chloride
Carbon disulfide	Acetone

Convulsants are usually considered a greater toxic hazard than depressants. They affect enzymatic pathways responsible for proper transmission of electric impulses from the brain, producing overstimulation of the nervous system and resulting in convulsions, muscle tremors, weakness of the lower extremities, and other abnormal sensations. Organophosphorous pesticides, such as parathion, are generally included in this group. Table 3.5 lists some central nervous system convulsants.

TABLE 3.5 Central Nervous System Convulsants

Fluoroacetate	Organophosphorous compounds
Hexachlorophene	Acrylamide
Lead	Thallium
Organomercury	Tellurium
Carbon disulfide	Triethyltin

The liver can be affected by toxic agents in three ways. Acute toxicity, or cell death, can result from exposure to a number of chemicals including carbon tetrachloride, chloroform, trichloroethylene, tetrachloroethane, bromobenzene, tannic acid, and kepone. Chronic toxicity, characterized by cirrhosis of the liver, is a progressive fibrotic condition induced by alcohol, carbon tetrachloride, and some heavy metals. The liver continually biotransforms most substances that enter the body, and in most cases, toxic agents are broken down into less harmful or inert products. In some instances, however, biotransformation may lead to the formation of toxic by-products. Carbon tetrachloride, for example, is broken down into chloroform by the liver. Although biotransformation in the liver eliminates the acute toxicity of xenobiotics, metabolic by-products may not be eliminated from the body and thus accumulate in the liver. In this case, chronic toxicity becomes a major concern.

The kidneys and the excretory system are susceptible to toxic chemical insult for several reasons. After the liver, the kidneys have the highest level of biotransformation activity. In addition, in a person at rest, 20–25% of the blood flow goes through the kidneys. The excretory system is characterized by active secretion processes that may concentrate toxicants in renal tissues. Heavy metals such as mercury, lead, and arsenic are readily concentrated in the kidneys and can lead to cell death. Organic compounds containing chlorine, fluorine, bromine, or iodine result in toxic metabolites after biotransformation in the kidneys.

3.5 LESS COMMON WORKPLACE TOXIC EFFECTS

The toxicological literature provides extensive coverage of other forms of toxicity not discussed here. Teratogenic effects are those that produce congenital malformations or abnormalities in the fetus. A few chemical agents, most notably organic mercury compounds and anesthetic gases, have been shown to produce these harmful effects in humans. Mutagens are chemicals that alter DNA, producing changes in the genetic code. These changes may include chromosomal aberrations, rearrangement of chromosome pieces, gain or loss of entire chromosomes, or DNA base changes within a gene. Examples of mutagenic chemicals are ethylene oxide, ethyleneamine, hydrogen peroxide, benzene, and hydrazine. Carcinogenic agents are those that lead to cancer. Common carcinogenic agents include asbestos, benzene, and vinyl chloride.

3.6 REFERENCES

1. U.S. EPA. *Hazardous Materials Incident Response Training Program*, 1983.
2. J. H. Meidl. *Explosive and Toxic Hazardous Materials*. Macmillan, New York, 1970.
3. N. I. Sax. *Dangerous Properties of Industrial Materials*. Van Nostrand Reinhold, New York, 1984.
4. F. A. Patty. *Industrial Hygiene and Toxicology*. John Wiley and Sons, New York, 1978.
5. NIOSH. *Registry of Toxic Effects of Chemical Substances*. Rockville, MD, 1980.
6. American Conference of Governmental Industrial Hygienists. *Threshold Limit Values and Biological Exposure Indices for 1985–86*, 1985.
7. M. Sittig. *Hazardous and Toxic Effects of Industrial Chemicals*. Noyes Data Corp., Park Ridge, NJ, 1979.
8. J. Doull, C. D. Kladsin, and M. O. Andur. *Toxicology, The Basic Science of Poisons*. Macmillan, New York, 1980.

4

Monitoring and Detecting Hazardous Chemicals

4.1 INTRODUCTION

The presence and concentration of hazardous substances at an incident scene or waste site must be monitored accurately and constantly so that response professionals as well as the general public are protected. Only when the nature and quantity of hazardous substances at a site are known can the levels of personal protection, as well as the areas where it is needed, be determined. Monitoring instrumentation allows responders to determine the effectiveness of their preventive or remedial actions and to evaluate the progress of the overall cleanup effort.

4.2 INSTRUMENT SELECTION CRITERIA

Field use of monitoring instruments is far different than instrument operation in the laboratory. At the site of an emergency, technical and structural demands are placed on instruments in excess of those observed under controlled conditions. It is necessary therefore to carefully evaluate all monitoring devices to be used at the scene of an emergency response. Common factors to consider when selecting field-deployable instrumentation include the following:

- **Portability**—Reinforced shells, shock-mounted electronic packages, padded shipping containers, weather-proof packaging, easily carried
- **Reliability**—Fast response time, immediately readable data, immediately interpretable data, consistent data
- **Selectivity and sensitivity**—Minimal interference, distinguishes substance of interest, accurate and repeatable at acceptable low and high concentrations
- **Inherent safety**—Electrical system constructed to eliminate arc from power source or electronics, control of flame or heat sources inherent in the instrument

Two terms often employed to communicate the safety of monitoring instruments used in ignitable atmospheres are *explosion-proof* and *intrinsically safe*. Explosion-proof devices encase all ignition sources in a rigidly constructed container. Should the instrument enter an area containing ignitable/explosive vapors, the resulting combustion would be contained within the instrument and all hot gases would be cooled prior to emission from the device. An intrinsically safe device is one incapable of releasing electrical or thermal energy sufficient to ignite volatile hazardous chemicals, under any usage conditions. Examples of harsh usage conditions included in this definition are

- Accidental damage to any wiring
- Failure of electrical components
- Application of over-voltage
- Adjustment and maintenance operations

Underwriters Laboratory (UL) and Factory Mutual Research Corporation (FM) provide testing and certification of many monitoring instruments under procedures established by the National Fire Protection Association (NFPA) and the American National Standards Institute (ANSI). These approvals are commonly sought by instrument manufacturers, and it is expected that instruments being considered bear one of these approvals. It is equally important to determine exactly what hazardous volatile gases or atmospheres are approved for operation of a particular instrument. For example, an instrument rated as intrinsically safe for an ethylene atmosphere would not necessarily be safe in a hydrogen atmosphere. The class, division, and group of hazardous or flammable atmospheres in which an instrument has been approved for usage must be marked on the instrument on a permanently affixed plate. Descriptions of these classes, divisions, and groups of hazardous or flammable atmospheres may be found in the National Electrical Code, published by the NFPA.

4.3 MONITORING INSTRUMENTS

4.3.1 Introduction

Monitoring instruments are useful in determining the presence and concentration of flammable atmospheres, oxygen, toxic vapors and gases, and ionizing radiation. These instruments are routinely used for initial surveys and incident or site evaluation, as well as for periodic or continuous monitoring throughout the work period.

4.3.2 Combustible Gas Indicators

Combustible gas indicators (CGIs) monitor combustible gases and vapors. CGIs typically detect and register the concentration of combustible gases and vapors as a percent of the lower flammable limit of the calibration gas. Most GCIs operate on the hot wire principle, where a filament, usually platinum, is heated and allowed to come in contact with combustible gases or vapors. The temperature of the filament normally increases when placed in the presence of combustible gases or vapors, thus increasing the electrical resistance of the filament. These changes in filament resistance are registered as the ratio of combustible vapor present to that required to reach 100% of the lower flammable limit.

CGIs should be calibrated using a hydrocarbon–air mixture that contains a hydrocarbon similar to that anticipated at the response scene. Lower flammable limit concentrations can vary widely between different hydrocarbons. An instrument calibrated for a 50% mixture of propane in air, for example, may not read precisely 50% if the mixture encountered is methane in air. Most alarms on CGIs are set at 10% of the lower flammable limit, so that errors resulting from encounters with flammable gases other than the calibration gas are minimized. Thus, for example, an alarm set for 10% of the lower flammable limit of propane could signal at 5% of the lower flammable limit for methane.

CGIs come in two classes:

- Those that require pumps to flow the sample gas over the sensor
- Those that rely on diffusion for sample input and analysis

CGI pumps may consist of a manual aspirator bulb or an automatic system supporting continuous air intake. Aspirator bulb instruments take in air samples discontinuously and can be used for periodic checks. Sensors that use diffusion-type cells have a slightly slower response time because atmospheric samples must pass through a semipermeable membrane before analysis. Diffusion-type apparatus are widely used as personal monitors for tank cleaning and confined space entry.

There are some important limitations to CGI technology:

- Differences between calibration temperature and sampling temperature can affect reading accuracy, unless the instrument is designed to compensate for this.
- Differences in physical and chemical properties between gases sampled and the calibration gas may lead to inaccurate readings.
- These systems can give erroneous readings when oxygen concentrations are abnormal.
- Certain substances can foul or destroy the filament, decreasing instrument sensitivity.
- Silicon-based compounds, often found in these instruments, can destroy filament integrity.

Advantages to CGI technology include the following:

- CGIs are easy to operate.
- Most CGIs have low battery alarms or indicators. However, these systems should be calibrated immediately prior to use in the field.

4.3.3 Oxygen Sensors

Oxygen sensors and CGIs are sometimes combined in the same instrument. Oxygen measurement can serve two purposes in emergency response situations:

- Oxygen determinations are necessary to determine CGI accuracy.
- Oxygen-level determinations indicate the presence of nonflammable gases or vapors that the CGI cannot detect.

Readings on these combined oxygen sensor/CGI instruments are commonly given as the percent oxygen concentration. In the immediate atmosphere, oxygen is normally expected to be about 20.9%. Commercial oxygen sensors found at emergency response sites traditionally provide readings from 0 to 25% oxygen concentration and provide warnings if the oxygen level drops below 19.5%. Some oxygen sensor units use pumps or aspirator bulbs to bring oxygen molecules through a membrane into an electrolytic solution; others rely on diffusion processes similar to the CGIs. Once in contact with the electrolytic solution, oxygen reacts with electrodes to generate a small electric current, which in turn produces a reading on the meter.

Oxygen sensors must be calibrated prior to usage to compensate for barometric pressure changes and changes in altitude, which affect normal oxygen concentrations. Carbon dioxide in high concentrations (7.5%) shortens the useful life span of the sensors once they are exposed to ambient air. Oxygen sensor units that allow field replacement of the oxygen cells provide greater instrument use and less down time.

4.3.4 Photoionization Detectors

Photoionization detectors (PIDs) are used to measure the concentration of organic and some inorganic vapors and gases on site. In PID units, hazardous substances are ionized with ultraviolet light producing an electrical current proportional to the number of ionized molecules generated. Because of the mechanism of detection, PIDs are much more sensitive than CGIs. PIDs can commonly record toxic chemical concentrations in the parts per million range and are thus one of the most useful devices for monitoring highly toxic organic substances at an emergency response site.

PIDs do have some significant drawbacks:

- They do not accurately detect methane.
- They do not detect many other toxic substances if the particular probe used at the time does not have a higher energy level than the ionization potential of the toxic substance being evaluated (the ionization potential is the energy in electron volts needed to cause ionization of a substance).
- Mixtures of gases cannot be measured quantitatively unless their ionization potentials are the same.
- Radio signals and other voltage sources can interfere with PID measurements.
- High humidity may affect PID response.

The effective use of a PID requires proper training in operation, maintenance, and data interpretation of the instrument. Although a PID may be capable of measuring concentrations of 0–2000 ppm of benzene, for example, the response may not be linear over this entire range. Above 600 ppm, the readings on the PID may be lower than the actual concentration. Training and familiarity with the instruments are therefore essential if the data are to be useful.

4.3.5 Flame Ionization Detectors

Like the PID, flame ionization detectors (FIDs) use ionization of contaminant molecules as the underlying means of analysis. However, in the FID, a hydrogen flame, rather than ultraviolet light, initiates chemical ionization. Flame ionization is conventionally useful in emergency response settings to detect organic gases in the parts per million concentration range. A conventional FID contains a pump that brings a gas sample into the detector chamber where the gas molecules contact a hydrogen flame. Most organic compounds burn easily, producing positively charged carbon ions; in a FID, these ions are collected, producing an electrical current proportional to their number.

Limitations of FID technology include the following:

- FID units do not detect inorganic gases or vapors efficiently.
- FID systems cannot ionize many synthetic organic substances.
- Sensitivity of FID units varies with the compound.
- At temperatures lower than 400°F, gases and vapors begin to condense in the FID pump and column, often causing inaccurate analytical responses.
- Oxygen-deficient atmospheres and high concentrations of contaminants require modifications to FID systems for long-term use in the field.
- FID units provide true readings only for the substances used in unit calibration. Readings of all other chemical agents must be adjusted to generate correct values for the compounds under evaluation.

Some FIDs can operate in a gas chromatograph (GC) mode as well as the survey mode. In the GC mode, hydrogen gas carries an injected sample of ambient air through a packed column. Different contaminants are retained on the column for varying lengths of time and are subsequently detected separately by the FID. A strip chart recorder graphically plots the retention times of the different contaminants, permitting comparison with known standards of various toxic chemical substances. This ability to operate in a dual GC/survey mode allows FID units to identify and/or measure specific organic agents in the field. However, highly experienced operators are needed when FIDs are used in the GC mode. The hydrogen fuel supply must be monitored, batteries should be checked regularly, and operating procedures should be carefully followed.

4.3.6 Infrared Spectrophotometers

Infrared (IR) spectrophotometers operate on the principle that the frequencies and intensity of IR radiation absorbed are specific for a compound and its concentration under particular environmental conditions. By measuring the frequency and quantity of IR radiation absorbed, measurements of parts per million concentrations of toxic substances can be achieved. Generally, IR spectrophotometers have limited field application at emergency response locations because of several factors:

- The exact identity of the hazardous contaminants present may not be known.
- The IR spectrophotometer has not been recognized as safe to operate in flammable conditions.
- Extensive experience is required for reliable IR analysis of toxic materials.

4.4 SPECIFIC CHEMICAL INSTRUMENTS

For a limited number of substances, specific detection instruments are commercially available. Most of these instruments use electrochemical cell technology based on principles of operation very similar to that described previously for oxygen sensors. For example, specific electrochemical sensors can be used for monitoring and detecting carbon monoxide, hydrogen sulfide, chlorine, hydrogen cyanide, and nitrogen dioxide under field conditions. Many manufacturers use one of these sensors in combination with a CGI and an oxygen sensor to construct a multi-purpose instrument. However, because interference from the often-complex mixture of toxic agents present at an emergency response scene is often reported as a problem for electrochemical instruments in the field, careful consideration should be given as to when this technology is useful on site.

4.5 COLORIMETRIC INDICATOR TUBE SAMPLERS

Colorimetric indicator tubes allow quick and easy measurement of specific gases and vapors. Tubes are available for a much wider range of chemical substances than specific electronic detection devices and are currently in wide use in industry. Colorimetric indicator tubes are made of glass and filled with a chemical reagent sensitive to the hazardous chemical agent of interest. To detect volatile hazardous substances, the ends of the tube are first broken off, and the tube is subsequently inserted into a manual bellows or piston-type pump. The pump is then used to draw a specific volume of air through the tube at a predetermined rate. If particular contaminants are present in the air sample, a color change or stain develops within the reagent inside the tube. The length of the stain is measured against a scale, usually marked on the glass tube, and the concentration is subsequently read in parts per million. Tubes are generally chemical specific, although some manufacturers have developed tubes that detect groups of chemicals. By employing a variety of tubes for different groups of hazardous substances, response professionals can narrow down the identity of an unknown compound.

Colorimetric indicator tubes as a field-deployable monitoring device for hazardous substances have some limitations:

- Sampling is not done in a continuous manner.
- Only a very small atmospheric region is actually sampled.
- Most tubes designed to detect one particular chemical agent also have a number of other hazardous substances that can interfere with detection of the targeted compound, providing false analytical readings.
- Determining the actual length of the stain on the tube is subjective, so that different users may report different values.
- Tubes have a limited shelf life (typically two years) and should be refrigerated to preserve reagents.

Different manufacturers use different reagents, tube sizes, and pump types to test for particular compounds. In standard usage, colorimetric indicator tubes are employed only to verify the presence of a toxic or hazardous substance and to determine the approximate concentration of a particular hazardous agent.

4.6 AIR MONITORING

Initial assessment and surveys of the emergency response site are traditionally used to set priorities for routine, ongoing air monitoring programs. Personnel should use extreme caution and wear appropriate personal protective equipment (PPE) when the following immediately dangerous to life and health (IDLH) conditions exist:

- Flammable atmospheres
- Oxygen deficiencies
- Toxic substances

Open areas generally receive the lowest monitoring priority because natural dispersal forces dilute atmospheric contaminants readily. Low-lying areas, confined spaces, and containers, however, commonly allow hazardous concentrations of substances to persist for extended periods of time and thus traditionally warrant higher monitoring priority. Air sampling must not be an ancillary part of emergency response actions and should be conducted with detailed, specific objectives identified as guiding principles to the responders. As an example, to identify and define the size of a contaminant plume, air sampling should begin downwind from the hazardous chemical source and continue along the axis of the wind until the source is reached or until safety becomes a factor. Next, cross-axis sampling should be performed to determine the width of the plume. Careful thought must be given to the air sampling plan undertaken so that accurate results can be achieved which properly define the extent of volatile hazardous agent contamination at the site.

Detection and quantification of unknown inorganic vapors and gases in the field prove difficult due to a lack of capable instrumentation. Colorimetric indicator tubes may have some application in this area, and PIDs can monitor a few inorganic materials. Unknown organic compounds can be quantified by PIDs and FIDs when used in the survey mode and identified to a limited degree when the GC mode is used. CGIs will detect the presence of many combustible gases and vapors at relatively high concentrations.

4.7 PERSONAL MONITORING

Personnel working in response areas where conditions might become hazardous due to unexpected events are often fitted with a personal monitoring instrument as a warning device to signal that a hazardous environment is present. In other cases, it is important to routinely monitor worker exposure to various known and suspected hazardous agents during normal work periods so as to obtain a time-weighted average (TWA) of the chemical exposures.

Personal monitoring instruments are used to detect a wide range of chemical hazards such as flammable vapors, low oxygen concentrations, and toxic gases. These devices must be small, compact, and easily carried by workers; audible as well as visual alarm signals are often used along with accessories such as special ear plug attachments if work is to be conducted in unusual environments such as situations with high noise. Other specifications often requested on commercial monitoring instruments include the following:

- A power supply, such as rechargeable batteries, that will last an entire work period
- Calibration adjustment knobs positioned or designed so they cannot be accidentally readjusted by workers

Generally, these instruments operate by much the same mechanisms described earlier (e.g., hot filament or wire mechanisms) but are far more compact. Microprocessor system attachments allow a TWA of worker exposure to be determined for specific hazardous contaminants of concern; this approach to personal monitoring also eliminates the need for sending large quantities of biological and air samples to centralized laboratories for analysis. More commonly, TWA exposures are determined by collecting environmental samples in an absorbent medium, followed by laboratory analyses at a later time to allow precise concentration determinations. Monitoring devices, including badges or tubes containing an absorbent, are worn by workers within the hazardous zone. Other systems employ a small microprocessor pump to draw a continuous flow of air through the sampler. Passive dosimeter badges rely on diffusion processes to bring contaminants into contact with the monitoring absorbent; some types of passive dosimeters can be read directly, as previously described for colorimetric tube sampling devices.

4.8 REFERENCES

1. J. B. Cumbus. *Continuous Monitoring of Air Quality with Portable Analyzers.* From a paper presented at the 1982 Industrial Safety Seminar, Texas Chemical Council.
2. U.S. EPA. *Hazardous Materials Incident Response Training Program*, 1983.

Industrial Hygiene Issues

5.1 INTRODUCTION

Uncontrolled hazardous chemical releases pose a number of risks to response professionals. Although the potential adverse health effects from hazardous chemical exposure are the primary concern for personnel working at an emergency release, numerous other hazards and risks characterize these working environments as well:

- Temperature extremes
- Noise exposure
- Ionizing radiation
- Oxygen deficiency
- Fire and explosion hazards
- Electrical hazards
- Unsafe conditions
- Biological hazards

This section describes some of these hazards commonly present at uncontrolled hazardous chemical release sites. Risk characterization of an emergency, which should be performed prior to response initiation, must include an inventory of these factors in the area to facilitate development of an accurate list of potential worker and responder hazards. A site safety and health program should subsequently be prepared to address all potential work hazards and provide adequate protection to all personnel entering and working at the location.

5.2 HEAT STRESS

Workers who must wear protective clothing are especially susceptible to heat stress because their bodies are insulated from the outside environment and cannot easily release heat. Personnel in the immediate working area are subjected to internal heat generated from exertion and external heat provided by the sun, fires, and equipment. When developing both a worker safety plan and a health monitoring program, three sources of heat exchange in the body must be considered:

- Conduction–convection
- Evaporation–convection
- Radiation

Conduction–convection takes place when air or fluid molecules contact the body and absorb as well as remove body heat. Evaporation–convection is a similar mechanism dependent on the capacity of the skin to produce sweat and subsequently release heat as water vaporizes from the body's surface. Air movement and a low atmospheric vapor content, in turn, greatly accelerate evaporation processes. Radiation is the transmission of energy by electromagnetic waves. The body, in this case, exchanges heat by transmitting infrared radiation.

The net change in body heat content is the sum of heat produced by metabolism and heat exchanged by the three processes just described. Metabolic heat can be calculated from the amount of oxygen used by the body (about 5 kcal (kilocalories) of heat is liberated for each liter of oxygen respired[1]). Special formulas, in addition, allow calculation of heat exchange by conduction–convection, radiation, and heat loss by evaporation–convection, including the following:

$$\Delta S = M - E + R - C$$

where ΔS = change in body heat content
M = metabolic heat produced by body
E = evaporative heat loss
R = radiant heat gain (or loss)
C = heat loss (or gain) by conduction–convection

Heat buildup can lead to a number of adverse health effects. Five major problems have been identified:

1. Heat stroke
2. Heat exhaustion
3. Heat cramps
4. Heat rash
5. Dehydration

Heat stroke results when the mean body temperature becomes intolerable for certain vital organs and tissues. Symptoms include hot, dry skin with a red, mottled, or bluish appearance; body temperature at 106°F and rising; and various brain dysfunction such as confusion, delirium, loss of consciousness, convulsions, and coma. Heat stroke has been known to occur without symptoms of heat exhaustion preceding it. A person suffering from heat stroke must have his or her body temperature lowered immediately by being wrapped in wet sheets or packed in ice.

Heat exhaustion results from an insufficient blood supply reaching the brain; symptoms include weakness, extreme fatigue, giddiness, nausea, and headache. Body temperature and sweating remain normal, but skin color may be pale or flushed. Dehydration in mild cases leads to a more rapid onset of heat exhaustion; in severe cases, muscular inefficiency, loss of appetite, reduced saliva, difficulty in swallowing, and nervous irritability may result.

Heat cramps are spasms of the voluntary muscles in response to low salt levels in the blood. The limbs and abdomen are the first muscles to evidence effects of low salt levels; this condition, nonetheless, may be prevented by increasing salt intake. Heat rash, another common workplace stress symptom, is the mildest of the heat problems resulting when sweat glands become inflamed.

Because the incidence of heat stress is dependent on a number of factors such as lack of physical fitness, age, acclimatization, infection, obesity, and drug use, all workers should be monitored even if they are not wearing protective clothing. The monitoring procedures suggested by NIOSH[2] are as follows:

- Monitor heart rate. Count radial pulse during a 30-sec period as early as possible in the rest period.
  a. If the heart rate exceeds 110 beats/min at the beginning of the rest period, shorten the next work cycle by one-third and keep the rest period the same.
  b. If the heart rate still exceeds 110 beats/min at the next rest cycle, shorten the following work cycle by one-third.
- Monitor oral temperature (clinical thermometer, 3 min under the tongue) at the end of the work period and before drinking.
  a. If oral temperature exceeds 99.6°F, shorten the next work cycle by one-third without changing the rest period.
  b. If oral temperature still exceeds 99.6°F at the beginning of the next rest period, shorten the following work cycle by one-third.
  c. Do not permit a worker to wear a semipermeable or impermeable ensemble if oral temperature exceeds 100.6°F.

- Monitor body water loss by measuring weight on a scale accurate to 0.25 lb at the beginning and end of each work day to see if enough fluids are being taken to prevent dehydration. Body water loss should not exceed 1.5% total body weight in a work day.

Acclimatization is a physiological phenomenon that permits adjustment of the human body to new or abnormal climatic conditions. It is induced when body core temperature and skin temperature are maintained at higher than normal levels for an hour or more per day. Most acclimatization occurs in the first two days of work and slowly progresses over a period of one to two weeks. However, some loss of acclimatization can occur in as little as two days, and after two weeks the loss is substantial.

Measures that can be taken to manage heat stress include adjusting work schedules, providing shelter or shaded rest areas, good maintenance of body fluids, good physical fitness, cooling devices under protective clothing, and worker training to recognize the signs and symptoms of heat stress. Table 5.1 summarizes the frequency of physiological monitoring for workers.[3]

TABLE 5.1 Frequency of Physiological Monitoring for Fit and Acclimatized Workers

Monitor after each ____ min of work under listed conditions:		
Adjusted Temperature	Normal Work Ensemble	Impermeable Ensemble
90.0°F or above	45 min	15 min
87.5–90.0°F	60 min	30 min
82.5–87.5°F	90 min	60 min
77.5–82.5°F	120 min	90 min
72.5–77.5°F	150 min	120 min

Source: NIOSH, OSHA, U.S. Coast Guard, U.S. EPA. *Occupational Safety and Health Guidance Manual for Hazardous Waste Site Activities.* DHHS Publication No. 85-115, 1985.

5.3 COLD STRESS

Persons working outdoors in temperatures at or below freezing are subject to frostbite in areas of the body with a high surface-area-to-volume ratio, such as fingers, toes, and ears. Two factors, ambient temperature and wind velocity, influence the development of cold injury. Wind chill, in addition, is a measure used to describe the chilling effect of moving air in combination with low temperature. Generally, the greatest wind chill effect results when a wind of 5 mph increases to 10 mph. A worker suddenly exposed to cold wind will experience very rapid body cooling.

Frostbite has several degrees of severity, which may be described as frost nip, superficial frostbite, and deep frostbite. Frost nip is characterized by a sudden blanching or whitening of the skin, superficial frostbite is characterized by waxy or white skin that is firm to the touch, and deep frostbite is extremely serious, resulting in cold, pale, and solid tissues.

Hypothermia results from exposure to rapidly dropping temperatures; symptoms are exhibited in five stages:

1. Shivering
2. Apathy, listlessness, and sometimes rapid cooling of the body to less than 95°F
3. Unconsciousness, glassy stare, slow pulse, and slow respiratory rate
4. Freezing of extremities
5. Death

5.4 NOISE EXPOSURE

Work conducted near equipment often leads to excessive worker noise exposure, which can cause hearing loss, annoyance, disruption of vital communications, and impaired performance. There are three basic types of noise:

- Wide band
- Narrow band
- Impulse

Wide band is sound distributed over many frequencies; narrow band is sound concentrated in a few frequencies, like the noise emitted from power tools; and impulse consists of sudden transient pulses such as that of a jackhammer or gunshot blasts.

Decibels (dB) are the common units for measuring relative sound levels and sound intensities. A decibel is approximately equal to the smallest degree of difference in loudness ordinarily detected by the human ear. The faintest audible sound to the human ear is 1 dB; consequently, 120 dB is the normal pain threshold for humans.[1] Frequency is the number of variations in sound pressure per unit time and is usually expressed as Hertz (Hz; cycles per second). Normal adults hear from between 20 and 20,000 Hz, depending on age (ability to hear in the mid to high range decreases with age).

TABLE 5.2 Allowable Noise Exposures in the Workplace

Noise Level (dBA)	Allowable Exposure (Hours)
90	8
95	4
100	2
105	1
110	0.5
115	0.25

Time-weighted average (TWA) exposure limits for noise have been established by the Occupational Safety and Health Administration (OSHA). Table 5.2 lists the OSHA permissible exposure limits for noise, measured in decibels on the "A" scale of a standard sound level meter (dBA). In addition, OSHA requires a hearing conservation program for employees with an 8-hr TWA exposure of 85 dBA.

5.5 IONIZING RADIATION

Radioactive materials emit one or more of three types of harmful radiation:

- Alpha particles
- Beta particles
- Gamma rays

Alpha particles have a limited ability to penetrate materials and are usually stopped by outer layers of skin and clothes. Alpha particles can create serious health problems, however, if they are ingested or inhaled. Beta radiation can produce surface "burns" to the skin and damage to the subsurface blood system; beta particles are also hazardous if inhaled or ingested. Good protection against alpha and beta radiation includes the use of personal protective clothing and respirators, along with careful decontamination and personal hygiene practices. Gamma radiation easily passes through clothing and the body tissues and can cause serious, permanent damage. Clothing for chemical protection will not prevent injury from gamma radiation; distance is the only safe recourse without special equipment. Radiation levels above 0.01–0.02 mrem/hr are considered above normal. If levels are detected at 2 mrem/hr or greater, all activities should cease until a physician has been consulted.

5.6 OXYGEN DEFICIENCY

At sea level, the oxygen content of air is approximately 20.9%. Environments with levels below 19.5% are considered oxygen deficient and should only be accessed with self-contained breathing apparatus or other similar breathing apparatus. The physiological effects of oxygen deficiency in humans become readily apparent when concentrations drop to about 16%; effects include impaired attention, judgment, and coordination and increased breathing and heart rate. At oxygen concentrations below 16%, nausea and vomiting, brain damage, heart damage, unconsciousness, and death can occur.

Oxygen deficiency may result from displacement of oxygen by another gas or the consumption of oxygen by some chemical reaction or biological process. Confined spaces and low-lying areas should be carefully monitored because they are sites where oxygen deficiency is frequently found.

5.7 FIRE AND EXPLOSION HAZARDS

Fires and explosions can occur due to conditions encountered at uncontrolled chemical release sites:

- Chemical reactions may generate heat, cause fires, and lead to explosions; flammable or explosive chemicals may be accidentally ignited.
- Oxygen enrichment can result in fire.
- Many chemical compounds are shock sensitive and explode easily.
- A number of materials kept under pressure may produce a violent container rupture if heated or if the container is suddenly breached.

Fires and explosions are generally the result of site activities such as moving or opening containers, accidentally mixing incompatible chemicals, or introducing an ignition source into a flammable atmosphere. When fires and explosions occur, responder personnel can face hazards from a number of sources. Intense heat may be produced, oxygen may be displaced or used up, flying objects may be thrown in many directions for considerable distances, and toxic materials may be released. Workers at hazardous chemical release sites must be reminded to constantly remain aware of these potential dangers.

5.8 ELECTRICAL HAZARDS

Workers and responders should be reminded to exercise caution when performing operations around or with electrical equipment. Overhead power lines, downed electrical wires, buried electrical cables, and power tools pose a danger from shock or electrocution. Low-voltage tools are always preferable in hazardous chemical environments; these tools ideally should be fitted with ground-fault interrupters and water-tight corrosion-resistant connecting cables whenever possible.

Fires and accumulation of excess heat are two other potential hazards encountered when using electrical equipment at an emergency response site. Care must be taken to monitor emergency personnel so that circuits are not overloaded and fuses are not installed with a higher than appropriate specified rating. Responders should further be reminded that all work activities should be suspended during electrical storms.

5.9 UNSAFE CONDITIONS

Unsafe working conditions are an inherent characteristic of sites contaminated by hazardous chemicals and of the work performed at these locations. Potential unsafe conditions include irregular terrain, slippery surfaces, and poor housekeeping. These site characteristics can cause personnel or large equipment to slip, trip, or fall. Precariously positioned objects, such as waste drums or tools, often fall, resulting in responder injury. Care must also be taken to monitor excavation activities to ensure that unstable conditions such as ditch collapse do not occur.

Protective equipment should always be carefully monitored to decrease unreasonable risks stemming from vision, movement, or communication impairment. Other equipment should be evaluated for intrinsic safety, defective features, and unguarded moving parts.

5.10 BIOLOGICAL HAZARDS

Medical, research, laboratory, and many other wastes often contain biological or infectious hazards. In some cases, biological hazards may naturally occur due to the location of an emergency or worksite. Infectious organisms or agents may be spread through the air and water at a site; microbial agents may also be encountered at considerable distances from their original source. Protective clothing and respiratory protection, however, can easily reduce exposure to these potential hazards.

Other possible sources of biological hazards include poisonous animals, insects, and plants. Certain animals, such as prairie dogs which can carry bubonic plague organisms or Rocky Mountain spotted fever, may be the host for other agents.

5.11 REFERENCES

1. Robert W. Allen, Michael D. Ells, and Andrew W. Hart. *Industrial Hygiene.* Prentice Hall, Englewood Cliffs, NJ, 1976.
2. NIOSH. *Hazard Evaluation Report.* TA-80-77-853. Chemical Control Corporation, Elizabeth, NJ, 1981.
3. NIOSH, OSHA, U.S. Coast Guard, U.S. EPA. *Occupational Safety and Health Guidance Manual for Hazardous Waste Site Activities.* DHHS Publication No. 85-115, 1985.
4. George D. Clayton and Florence E. Clayton, Eds. *General Principles.* Vol. 1 in *Patty's Industrial Hygiene and Toxicology,* 3rd ed. John Wiley and Sons, New York, 1978.

6

Medical Surveillance Programs

6.1 INTRODUCTION

Emergency response personnel and hazardous waste site workers may be exposed to a variety of hazards including toxic chemicals, fires, and explosions. U.S. Occupational Safety and Health Administration (OSHA) regulations found in 29 CFR 1910.120 mandate the establishment of a medical surveillance program for these workers, and 29 CFR 1910.134 outlines several medical requirements for personnel wearing respirators.

By screening potential employees or response team members for their health and fitness, a good medical surveillance program can accomplish a number of objectives, such as establishing worker suitability for particular jobs. In addition, in an emergency, proper treatment can be administered in a timely fashion. Abnormal health symptoms are more easily detected if the general health and well-being of employees is periodically monitored. Furthermore, accurate medical surveillance records can serve as evidence in legal cases, assist in epidemiological studies, and provide precise reports to federal, state, and local agencies as required.

The effectiveness of a medical program depends greatly on active worker involvement. Management encouragement is therefore recommended, and it can take several forms including promotion of good health practices (i.e., physical fitness, avoidance of tobacco), assurance of confidentiality, and an indication that any suspected exposure or unusual symptoms will be reported.

6.2 PROGRAM DEVELOPMENT

The following components should be included in a medical program:

- Surveillance
  1. Pre-employment
  2. Periodic medical exams and follow-up exams when appropriate
  3. Termination examination
- Treatment
  1. Emergency
  2. Nonemergency
- Recordkeeping
- Program review

Medical programs should be specific, addressing particular personnel needs and the potential for exposure. They should be designed by a qualified occupational health physician or consultant, who should receive input from the site safety officer. When appointing a program manager, preference should be given to a physician with board certification in occupational medicine or extensive experience in this area. Laboratories chosen to perform medical tests should meet minimum requirements outlined in 42 CFR Part 74 Subpart M Section 263 [a] or meet conditions for coverage under Medicare. Table 6.1 describes the recommended elements for an effective medical program.

6.3 SURVEILLANCE

Examinations under OSHA regulations 29 CFR 1910.120 must be provided for employees who (1) are routinely exposed to hazardous substances (defined as 30 days or more in a year), (2) wear a respirator for 30 days or more in a year, or (3) may be engaged in emergency control of hazardous materials. An examination must also be provided for any employee suspected of being exposed to hazardous materials above the permissible exposure limits in an emergency situation.

Medical surveillance should be arranged before an employee is assigned to a particular task and repeated on an annual basis thereafter. A medical exam is also mandatory when an employee is terminated or reassigned unless the employee has had an exam within the last six months. Further, a medical exam is required when an employee reports signs or symptoms of possible overexposure. A physician can recommend more frequent exams for particular employees.

Employers must provide examining physicians with a copy of 29 CFR 1910.120 and the following information:

TABLE 6.1 Recommended Medical Program

1. Pre-Employment Screening
   - Medical history
   - Occupational history
   - Physical examination
   - Protective equipment fitness determination
   - Baseline monitoring for specific exposures
2. Periodic Medical Estimations
   - Yearly update of medical and occupational history
   - Yearly physical examination; testing based on (a) examination results, (b) exposures, (c) job class and task
   - More frequent testing based on specific exposures
3. Emergency Treatment
   - Provide emergency first aid on site
   - Develop liaison with local hospital and medical specialists
   - Arrange for decontamination of victims
   - Arrange in advance for transportation of victims
   - Transfer medical records; give details of incident and medical history to next care provider
4. Nonemergency Treatment
   - Develop mechanism for nonemergency health care
5. Recordkeeping and Review
   - Maintain and provide access to medical records in accordance with OSHA and state regulations
   - Report and record occupational injuries and illnesses
   - Review the site safety plan regularly to determine if additional testing is needed
   - Review the program periodically; focus on current site hazards, exposures and industrial hygiene standards

- The physical requirements of an employee's job
- Symptoms of exposure to substances being handled
- Employee duties
- Anticipated exposure levels
- A description of personal protective equipment (PPE) used
- Information from previous medical exams

Written copies of exam results, tests done, and medical diagnosis must be provided to employer and employee. In addition, both parties should receive written recommendations from the examining physician regarding limited use of respirators and other types of PPE and a statement outlining employee awareness of special conditions. Examination costs and regular employee wages for the time spent during examination must be borne by the employer. Records of all medical information given or received under this rule must be maintained by the employer for 30 years.

6.4 PRE-EMPLOYMENT SCREENING

The three major objectives of this program component are to:

- Ensure employee fitness for specific jobs
- Ensure the employee can work under the conditions imposed (i.e., use of PPE and exposure to certain chemicals)
- Provide a baseline record for comparison with future medical data

Good pre-employment screening with regard to occupational and medical history can help determine past illnesses, potential allergic reactions to certain chemicals, and medical problems related to lifestyle such as the use of tobacco, alcohol, and drugs.

A complete physical should include an examination of all body organs, especially the respiratory, cardiovascular, and musculoskeletal systems. Obesity, poor physical conditioning, and other problems that could hasten heat stroke should be noted. Conditions that could adversely affect PPE use (i.e., respirators) and effectiveness include perforated eardrums, facial scars, dentures, poor eyesight, and missing or otherwise dysfunctional fingers.

The ability to work while wearing PPE is a key factor evaluated during pre-employment screening. Any worker limitations concerning PPE use should be recorded. Screening should point out persons with a history of severe lung disease, heart disease, back problems, or orthopedic problems. Additional tests such as chest x-rays, pulmonary function testing, and electrocardiograms may be needed to determine the full extent of an individual's medical limitations.

The establishment of baseline medical data is a key factor of pre-employment screening. This information can verify the effectiveness of PPE and engineering controls, and it can correlate adverse health effects with exposure to hazardous chemicals.

Table 6.2 lists tests frequently performed by occupational physicians that can help provide baseline medical data. Typically, a series of medical screening and biologic monitoring tests should be run based on past occupational history and anticipated types of exposure. Appendix 6.1 provides a general sample of what a pre-employment examination should include.

TABLE 6.2 Tests Frequently Performed By Occupational Physicians

Function	Test	Example
Liver		
General	Blood test	Total protein, albumin, globulin; direct bilirubin if total is elevated
Obstruction	Enzyme test	Alkaline phophatase
Cell injury	Enzyme tests	Serum glutamicoxaloacetic transaminase (SGOT), serum glutamic pyruvic transaminase (SGPT)
Kidney: general	Blood tests	Blood urea nitrogen (BUN), creatinin, uric acid
Multiple systems and organs	Urinalysis	Including color, appearance, specific gravity, pH qualitative glucose, bile
Blood-forming function	Blood test	Complete blood count (CBC)

6.5 PERIODIC MEDICAL AND TERMINATION EXAMS

Periodic examinations can prove useful complements to information obtained during pre-employment screening examinations by allowing for comparisons and detection of potential adverse trends. The basic periodic exam should include the same elements as pre-employment screening, but should focus on any changes in type or amount of exposure, reported symptoms, or other specific factors that may have altered an individual's general health. The physician performing periodic examinations should be provided with information regarding worker exposure subsequent to pre-employment screening. A battery of tests for a range of potential exposures is generally not necessary, and physicians may choose to conduct tests only for possible exposure to specific agents. Hearing and visual checks as well as a general physical examination are also recommended as part of annual examinations.

Termination exams may be limited to obtaining the medical history, laboratory tests, and physical examination results of the last periodic examination if the following criteria are met:

- The last full medical exam occurred within the last six months
- No exposure occurred since the last examination
- No symptoms associated with exposure occurred since the last examination

If any one of these criteria is not met, a full examination is required and follows the same format as the sample pre-employment examination (Appendix 6.1).

6.6 EMERGENCY MEDICAL TREATMENT

In emergency response and in hazardous waste operations, each site has specific and potential hazards that must be addressed as part of overall emergency operation plans. Chemical, physical, and biological hazards should be evaluated and planned for, with special emphasis on emergency medical treatment. Plans should include response personnel, site workers, contractors, visitors, and outside response groups and agencies. Recommended guidelines for establishing an emergency treatment program include the following:

- A team of site personnel should be trained in first aid, including a certification in CPR. This first aid training should emphasize treatment of acute chemical toxicity, victims of fires and explosions, and heat stress. An emergency medical technician should be included on this team, when possible.
- Personnel should receive specific training in emergency decontamination as part of the overall emergency response plan.
- Personnel responsibilities and duties should be preassigned and practiced in response drills and exercises.
- An emergency first aid station should be established so that patients requiring off-site treatment can be stabilized and minor injuries can be treated by general first aid. Standard first aid equipment items such as an emergency eyewash, deluge shower, stretchers, decontamination solutions, ice, and portable water should be included.
- Medical specialists should be available for emergency consultations, including at least one physician on call 24 hours a day. These specialists may require information about specific chemical hazardous effects, symptoms, antidotes, and treatments.
- Standard procedures for monitoring workers for heat stress should be implemented (Table 6.3).
- Plans should be made in advance to provide for emergency transportation and contamination control for nearby medical facilities. Proper preplanning and training of ambulance attendants as well as hospital staff members are critical to emergency operations; thus, practice drills may need to be developed.
- Names, phone numbers, addresses, and procedures for contracting outside assistance should be conspicuously posted. These may include physicians, medical specialists, ambulance services, medical facilities, emergency service agencies, and poison control hotlines.

TABLE 6.3 Signs and Symptoms of Heat Stress

Heat rash may result from continuous exposure to heat or humid air.

Heat cramps are caused by heavy sweating with inadequate electrolyte replacement. Signs and symptoms include:

- Pale, cool, moist skin
- Heavy sweating
- Dizziness
- Nausea
- Fainting

Heat stroke is the most serious form of heat stress. Temperature regulation fails and the body temperature rises to critical levels. Immediate action must be taken to cool the body before serious injury and death occur, and competent medical help must be obtained. Signs and symptoms of heat stroke include:

- Red, hot, unusually dry skin
- Lack of reduced perspiration
- Nausea
- Dizziness and confusion
- Strong, rapid pulse
- Coma

- Site maps and directions to medical facilities should be provided and should be familiar to medical response personnel and managers.
- A good communications network should be established, including on-site radio communications.
- Emergency medical procedures should be reviewed with all arriving response personnel and at safety meetings prior to each work shift.

6.7 NONEMERGENCY MEDICAL TREATMENT

Exposure to hazardous substances may lead to adverse health effects or symptoms that require treatment and evaluation but are not necessarily considered life threatening. Plans should be made to ensure that any potential job-related symptoms of exposure are evaluated. Common illnesses such as allergies, colds, and flus which could impair workers who use respirators or other protective equipment may also merit evaluation.

6.8 MEDICAL RECORDS

Good medical recordkeeping is vital. Chronic effects may not show up for many years and not until after workers have responded to many incidents or worked at many hazardous waste sites. Good records, therefore, ensure that the worker's history of chemical exposure is on file and available throughout his or her career.

Unless specific standards specify otherwise, OSHA requires the employer to maintain and preserve medical records on exposed workers for years after they leave employment (29 CFR 1910.120). Results of medical testing and full medical records and analyses must be made available to workers, their authorized representatives, and authorized OSHA representatives. Records of all occupational illnesses and injuries must be maintained and posted yearly in a summary report (29 CFR 1904).

6.9 PROGRAM REVIEW

The continued effectiveness of any program, including a medical program, depends on ongoing review, evaluation, and upgrading. The site safety officer, medical consultant, and/or management representative should do each of the following at least once annually:

- Ascertain that each accident or illness was investigated promptly to determine its cause and that necessary health and safety procedural changes were made
- Evaluate the medical testing program to ensure that it is properly monitoring and screening for potential site exposures
- Add or delete medical tests as suggested by current industrial hygiene and other data
- Review potential exposures and site safety plans at all sites to determine if additional testing is required
- Review emergency treatment procedures and update emergency numbers and contacts

6.10 REFERENCES

1. NIOSH, OSHA, U.S. Coast Guard, U.S. EPA. *Occupational Safety and Health Guidance Manual for Hazardous Waste Site Activities.* DHHS Publication No. 85-115, 1985.
2. OSHA. *Hazardous Waste Operations and Emergency Response.* 29 CFR 1910.120, August 10, 1987.

3. Marc K. Shay. *Hazardous Waste Workers Health and Safety Training—Section 126 SARA Requirements.* Presented at the Indiana Hazardous Waste Conference, Anderson, Indiana, June 2, 1987.

APPENDIX 6.1
SAMPLE PRE-EMPLOYMENT EXAMINATION

6.A.1 Occupational and Medical History

Do a complete medical history emphasizing the following: nervous system, skin, lungs, blood-forming system, cardiovascular system, gastrointestinal system, genitourinary system, reproductive system, and the ears, nose, and throat.

6.A.2 Physical Examination

Include at least the following:

- Height, weight, temperature, pulse, respiration, and blood pressure
- Head, nose, and throat
- Eyes—Include vision tests that measure refraction, depth perception, and color vision. These tests should be administered by a qualified technician or physician. Vision quality is essential to safety, the accurate reading of instruments and labels, the avoidance of physical hazards, and for appropriate response to color-coded labels and signals.
- Ears—Include audiometric tests, performed at 500, 1000, 2000, 3000, 4000, and 6000 Hz pure tone in an approved booth (see requirements listed in 29 CFR 1910.95, Appendix D). Tests should be administered by a qualified technician and results read by a certified audiologist or a physician familiar with audiometric evaluation. The integrity of the eardrum should be established because perforated eardrums can provide a route of entry for chemicals into the body. The physician evaluating employees with perforated eardrums should consider the environmental conditions of the job and discuss possible specific safety controls with the site safety officer, industrial hygienist, and/or other health professionals before deciding whether such individuals can safely work on site.
- Chest (heart and lungs)
- Peripheral vascular system
- Abdomen and rectum (including hernia exam)
- Spine and other components of the musculoskeletal system

- Genitourinary system
- Skin
- Nervous system

6.A.3 Tests

- Blood
- Urine
- A 14 × 17-in. posterior/anterior view chest x-ray with lateral or oblique views only if indicated or if mandated by state regulations. The x-ray should be taken by a certified radiology technician and interpreted by a board-certified or board-eligible radiologist. Chest x-rays should not be repeated more than once a year, unless otherwise determined by the examining physician.

6.A.4 Ability to Perform While Wearing Protective Equipment

To determine a worker's capacity to perform while wearing protective equipment, additional tests may be necessary, for example:

- Pulmonary function testing—Measurement should include forced expiratory volume in 1 sec (FEV_1), forced vital capacity (FVC), and FEV_1-to-FVC ratio, with interpretation and comparison with normal predicted values corrected for age, height, race, and sex. Other factors such as forced expiratory flow, maximum expiratory flow rate, maximum voluntary ventilation, functional residual capacity, residual volume, and total lung capacity may be included for additional information. A permanent record of flow curves should be conducted by a certified technician and the results interpreted by a physician.
- Electrocardiogram (EKG)—At least one standard, 12-lead resting EKG should be performed at the discretion of the physician. A "stress test" (graded exercise) may be administered at the discretion of the examining physician, particularly for personnel working in conditions where heat stress may occur.

6.A.5 Baseline Monitoring

If there is a likelihood of potential on-site exposure to a particular toxicant, specific baseline monitoring should be performed to establish data relating to that toxicant.

Respiratory Protection: Principles and Practices

7.1 INTRODUCTION

The respiratory system is the primary target site for the action of many toxic agents and is the major route by which many harmful substances enter the bloodstream and subsequently damage specific organs and tissues. Engineering controls in some instances may be used to maintain a safe atmosphere for workers, yet in other cases various types of respirators may be needed to provide adequate respiratory system protection.

7.2 EQUIPMENT SELECTION

There are two basic types of respiratory protective devices:

- Atmosphere-supplying respirators
- Air-purifying respirators

There are advantages and disadvantages to each system as well as selection criteria and principles of operation.

Atmosphere-supplying respirators may be classified as self-contained breathing apparatus (SCBAs) or supplied-air respirators (SARs). SCBAs provide air from a source carried by the user, and SARs provide air via an air line from a remote source. Air-purifying respirators use ambient air, which passes through a filter or purifying material that absorbs contaminants before the air reaches the user.

These three classes of respirators can be further differentiated by the type of air flow that is supplied to the facepiece. The pressure inside the facepiece of a positive-pressure respirator is maintained higher than outside atmospheric pressure—during inhalation and exhalation. This positive pressure may be maintained by supplying a continuous flow of air through the facepiece or by a system referred to as pressure demand. Continuous-flow systems use large air volumes because a constant flow of air must enter and exit the facepiece. Pressure-demand systems use a pressure regulator and an exhalation valve to maintain constant pressure. Negative-pressure respirators simply rely on the negative pressure created by the user's inhalation to bring air into the facepiece. If leaks develop in this type of system, however, outside contaminants can be drawn into the facepiece through inhalation.

Each type of respirator can be assigned a protection factor (PF), which is a numerical rating of its ability to prevent contaminants from entering the facepiece. The PF is determined experimentally and is useful in making decisions regarding the types of respiratory protection needed when contaminants are known and their concentrations have been measured. A PF of 100 indicates that a worker can theoretically enter an atmosphere of up to 100 times the permissible exposure limit (PEL) with that particular type of respiratory device. PELs follow OSHA exposure guidelines, which in most cases are identical to threshold limit value (TLV) designations. ANSI publishes a list of PFs along with other information on respiratory protection in *American National Practices for Respiratory Protection.* Table 7.1 provides the ANSI list.

7.3 SELF-CONTAINED BREATHING APPARATUS

When potential contaminants are unknown or concentrations are at or above the immediately dangerous to life and health (IDLH) level, air-purifying respirators should not be used. SCBAs are often chosen in these situations over SARs because workers are not confined by an air line.

There are two types of SCBA: (1) open circuit and (2) closed circuit. Open-circuit systems use a cylinder of compressed air, whereas closed-circuit systems use a small cylinder or other source of pure oxygen. Various types of SCBAs can supply breathable air for periods of time ranging from 5 min to 4 hr. Only positive-pressure SCBAs are recommended when concentrations of contaminants are above the IDLH. SCBA units can be bulky and heavy, possibly contributing to heat stress and impairing movement in close quarters. Table 7.2 provides a comparison of various types of SCBAs.

TABLE 7.1 Respirator Protection Factors

	Facepiece Pressure	Protection Factor
I. Air-purifying respirators		
A. Particulate removing		
Single-use dust	–	5
Quarter-mask dust	–	5
Half-mask dust	–	10
Half- or quarter-mask fume	–	10
Half- or quarter-mask high efficiency	–	10
Full facepiece high efficiency	–	50
Powered high efficiency all enclosures	+	1,000
Powered dust or fume all enclosures	+	X
B. Gas and vapor removing		
Half mask	–	10
Full facepiece	–	50
II. Atmosphere supplying		
A. Supplied air		
Demand half mask	–	10
Demand full facepiece	–	50
Hose mask without blower full facepiece	–	50
Pressure-demand half mask	+	1,000
Pressure-demand full facepiece	+	2,000
Hose mask with blower full facepiece	–	50
Continuous-flow half mask	+	1,000
Continuous-flow full facepiece	+	2,000
Continuous-flow hood helmet or suit	+	2,000
B. Self-contained breathing apparatus		
Open-circuit demand full facepiece	–	50
Open-circuit pressure-demand full facepiece	+	10,000
Closed-circuit O_2 tank-type full facepiece	–	50
III. Combination respirator		
A. Any combination of air-purifying and atmosphere-supplying respirator	Use minimum protection factor listed above for type and mode of operation	
B. Any combination of supplied-air respirator and a self-contained breathing apparatus		

TABLE 7.2 Types of Self-Contained Breathing Apparatus (SCBA)

Type	Description	Advantages	Disadvantages	Comments
Entry and escape SCBA Open-circuit SCBA	Supplies clean air to the wearer from a cylinder. Wearer exhales air directly to the atmosphere.	Operated in a positive-pressure mode, open-circuit SCBAs provide the highest respiratory protection currently available. A warning alarm signals when only 20–25% of the air supply remains.	Shorter operating time (30–60 min) and heavier weight (up to 35 lb [13.6 g]) than a closed-circuit SCBA.	The 30–60 min operating time may vary depending on the size of the air tank and the work rate of the individual.
Closed-circuit (rebreather) SCBA	These devices recycle exhaled gases (CO_2, O_2, and nitrogen) by removing CO_2 with an alkaline scrubber and replenishing the consumed oxygen with oxygen from a liquid or gaseous source.	Long operating time (up to 4 hr) and lighter weight (21–30 lb [9.5–13.6 kg]) than open-circuit apparatus. A warning alarm signals when only 20–25% of the oxygen supply remains. Oxygen supply is depleted before the CO_2 sorbent scrubber supply, thereby protecting the wearer from CO_2 breakthrough.	At very cold temperatures, scrubber efficiency may be reduced and CO_2 breakthrough may occur. Units retain the heat normally exchanged in exhalation and generate heat in the CO_2 scrubbing operations, adding to the danger of heat stress. Auxiliary cooling devices may be required. When worn outside an encapsulating suit, the breathing bag may be permeated by chemicals, contaminating the breathing apparatus and the respirable air. Decontamination of the breathing bag may be difficult.	Positive-pressure closed-circuit SCBAs offer substantially more protection than negative-pressure units, which are not recommended on hazardous waste sites. While these devices may be certified as closed-circuit SCBAs, NIOSH cannot certify closed-circuit SCBAs as positive-pressure devices due to limitations in certification procedures currently defined in 30 CFR Part 11.

Escape-only SCBA	Supplies clean air to the wearer from either an air cylinder or from an oxygen-generating chemical. Approved for escape purposes only.	Lightweight (10 lb [4.5 kg] or less), low bulk, easy to carry. Available in pressure-demand and continuous-flow modes.	Cannot be used for entry.	Provides only 5–15 min of respiratory protection, depending on the model and wearer breathing rate.

The following are key questions to ask when assessing the need for and feasibility of a SCBA:[1]

- Is the atmosphere IDLH or likely to become IDLH? If yes, then positive-pressure SCBA should be used.
- Is the duration of the air supply sufficient for accomplishing the necessary tasks? If not, available options include choosing a different type of respirator, increasing the amount of air carried, or changing the work plan.
- Will the bulk and weight of the SCBA interfere with the tasks to be performed or cause added stress for the worker? If yes, shorten the work until weather conditions improve.

Open-circuit SCBAs use a cylinder of compressed air as the source of oxygen for the user. Exhaled air is exhausted from the system and not recycled. The user carries a full supply of air in the cylinder on his or her back; the supply is therefore limited by the bulk and weight a user can be expected to carry. Available units can last from 5 to 60 min; units of 5–15 min duration are approved as "escape only" and should not be used for entry.

Air used in open-circuit SCBAs must meet requirements set forth in the Compressed Gas Association's Pamphlet G-7.1. This requirement calls for at least "Grade D" air, which contains 19.5–23.5% oxygen and predominantly nitrogen. It also specifies a limit on condensed hydrocarbons, carbon monoxide, and carbon dioxide and prohibits undesirable odors.

The components of an open-circuit SCBA include the cylinder, high-pressure hose, regulator assembly, breathing hose and facepiece, and backpack and harness assembly. Figure 7.1 illustrates these features.

The cylinder of compressed air is considered a hazardous material and must meet certain specifications according to the Department of Transportation (DOT). A hydrostatic test must be performed on steel and aluminum cylinders every five years, and composite cylinders require testing every three years. In addition, the gauge on the cylinder must be accurate to ±5%.

The cylinder and the regulator are connected by a high-pressure hose that must be tightened by hand, never with a wrench. An O-ring provides the seal. The regulator receives air that passes from the cylinder through the high-pressure hose. In the regulator, air can take one of two routes. The bypass valve, when opened, allows air to flow at full pressure to the facepiece. With the bypass closed and the mainline valve open, air passes into the regulator and the pressure is reduced to 50–100 psi. A pressure relief valve is located after the pressure reducer in case of a malfunction, and a pressure gauge monitors the air pressure coming into the regulator before it enters the valves. It must be accurate to ±5% and should be monitored by the user to determine the available supply of air. Air leaving the reducing valve next encounters the admission valve, which in posi-

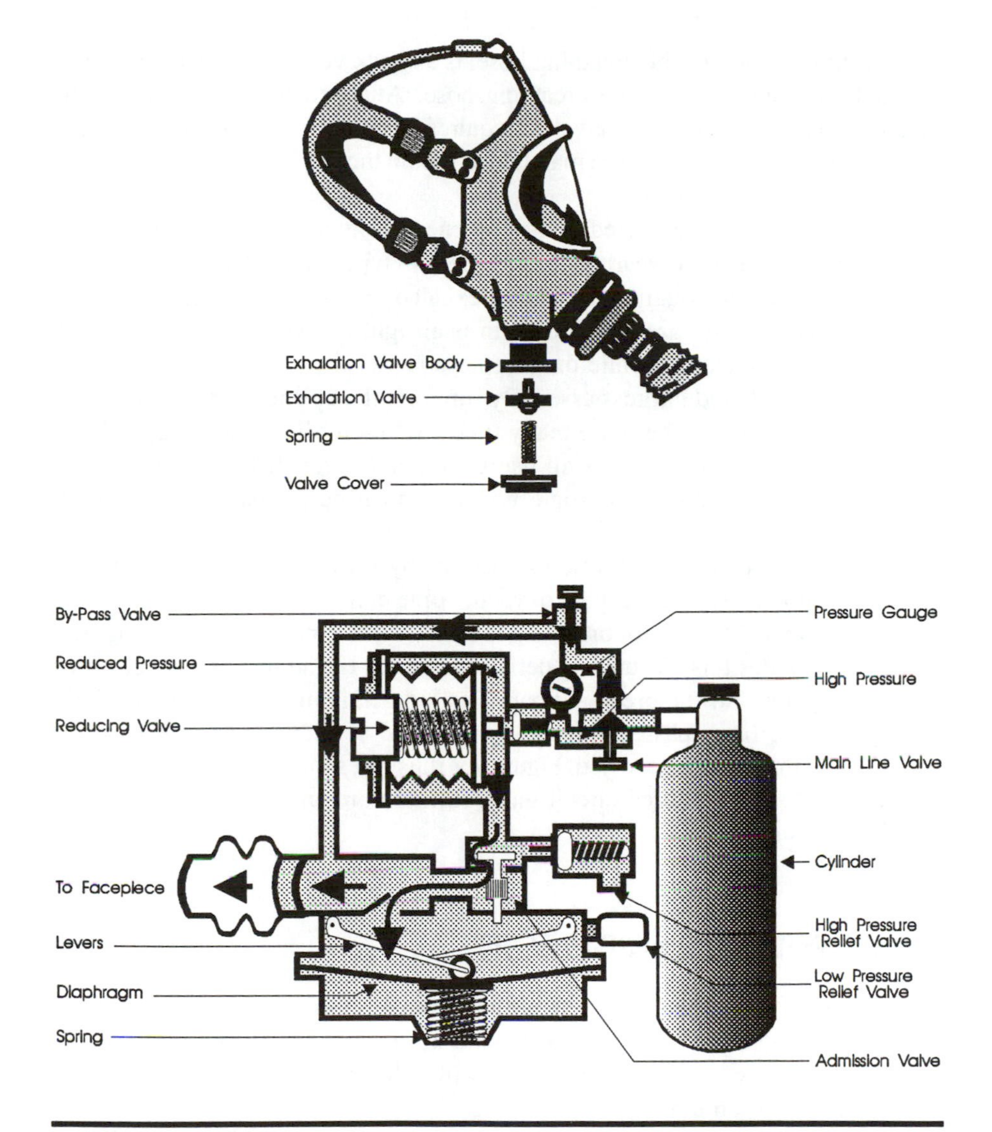

FIGURE 7.1 Open-Circuit SCBA

tive-pressure units remains open until enough back pressure is built up to close it. Back pressure builds because air is prevented from leaving the system until the user exhales. The admission valve must close at a pressure lower than that required to open the exhalation valve.

The regulator is connected to the facepiece by the breathing hose. This hose has rubber gaskets in the connectors at both ends for a good seal, and it should be corrugated or of sufficient length to allow flexing. Inside the facepiece, above the attachment point of the breathing hose, is a check valve. This valve prevents exhaled air from entering the breathing hose. Attached to the bottom of the facepiece is the exhalation valve. It requires about 2–3 in. of water column pressure to open (static pressure maintained within the facepiece is around 1.5 in. of water column pressure).

The facepiece is constructed from neoprene, or sometimes silicon rubber, and it typically has a clear polycarbonate lens. Certain types of mask have a nose cup, which helps prevent fogging and allows exhaled air to pass directly out of the facepiece. Some masks are also fitted with an airtight speaking diaphragm, which assists in communication while the facepiece is worn.

The backpack and harness assembly supports the cylinder and attaches the SCBA to the wearer. There are many types of assemblies available, and one model may not be suitable for all users. Appendix 7.1 describes inspection and checkout procedures for some types of positive-pressure, open-circuit SCBAs.

Closed-circuit SCBAs recycle exhaled air by scrubbing its carbon dioxide and mixing it with a supply of pure oxygen. Some units are capable of supporting the user for up to 4 hr. Air for breathing is mixed in an expandable breathing bag; as the user inhales this air, the bag deflates, opening the admission valve, which releases oxygen into the breathing bag. The oxygen is mixed with exhaled air, which has just been passed through an alkaline scrubber to remove carbon dioxide, and the cycle is repeated. Figure 7.2 illustrates the operation of a closed-circuit SCBA. Some closed-circuit units, furthermore, are designed to provide positive pressure.

7.4 SUPPLIED-AIR RESPIRATORS

SARs provide air to the user via an air line from a remote air supply. SARs can use either negative or positive pressure and always use air, never pure oxygen. A worker entering an IDLH atmosphere with a SAR must also wear an escape SCBA in case of an emergency.

Cylinders of compressed Grade D air or a compressor with the ability to purify air may be used as the air supply. OSHA has set standards for compressor use in 29 CFR 1910.134 [d]. SARs acceptable for use with compressors are designated "Type C" by the Mine Safety and Health Administration (MSHA) and NIOSH. The use of compressors at emergency scenes and hazardous waste sites should be carefully evaluated to ensure that the quality of the air is not compromised by the presence of hazardous contaminants.

SARs have an advantage over SCBAs in that they are lighter and less bulky. Dragging extended lengths of hose, up to 300 ft, can be difficult, however, and

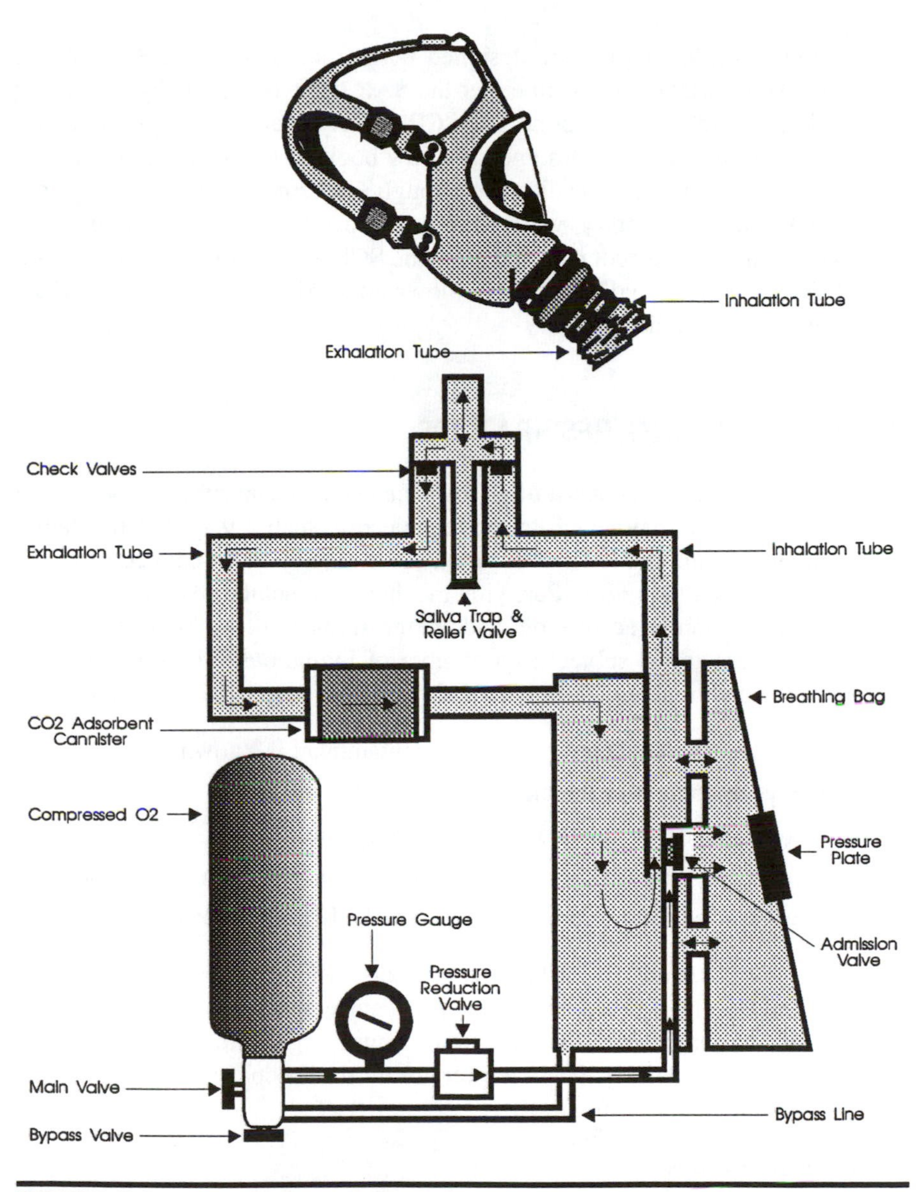

FIGURE 7.2 Closed-Circuit SCBA

tends to impair mobility. When leaving an area, the user must retrace his or her steps to avoid entangling the air line. Air lines are subject to punctures, burn-through, permeation by contaminants, and other problems that could affect the user's air supply. Vehicles and other workers should stay away from personnel using SARs.

New units on the market are designed with a combination of SAR/SCBA systems. The user may operate in either the SAR or SCBA mode by manual or automatic switching of air supplies. The SCBA is a typical positive-pressure, 30- or 60-min unit that has been designed to allow hook-up to an air-line system. A worker who has to travel some distance through a hot zone could use this system in the SCBA mode for entry, work for an extended time by connecting to the supplied-air line, then disconnect and go to the SCBA mode for his or her return. The NIOSH certification of these units allows up to 20% of the available SCBA air supply to be used during entry.

7.5 AIR-PURIFYING RESPIRATORS

Air-purifying respirators consist of a facepiece and an air-purifying device. The air-purifying device may attach to the facepiece in such a way that inhalation brings in outside air, or the air-purifying device may be worn connected to the facepiece by a breathing hose. Purifying mechanisms include several processes: filtration, absorption or adsorption, and chemical reaction. Air purification is a selective process and is subject to a number of limitations. The following are some of the conditions under which an air-purifying respirator may be used:

- The identity and concentration of the contaminant is known
- The oxygen content in the air is at least 19.5%
- The contaminant has adequate warning properties
- Approved canisters for the contaminant and concentration are available
- Contaminant concentrations do not exceed IDLH conditions

Except for powered air-purifying respirators, which use a battery-powered pump to draw air through the air-purification device, most air-purifying respirators operate by negative pressure. Powered air-purifying respirators send purified air to the user, maintaining positive pressure in the facepiece. There are three types of air-purifying respirators:

- Particulate filters
- Cartridges and canisters that contain sorbent medium
- Combination devices that contain layers of sorbent materials or sorbent material and filters

Combination devices may be useful for more than one contaminant, but they have not worked efficiently when tested for effectiveness against simultaneous exposures to more than one agent.

Particulate filters are designed to remove potentially harmful dusts, fumes, or mists. They are classified according to their effectiveness at removing particles of various sizes. Mechanical filters can load up, making breathing more difficult. Sorbent filters are designed to remove specific contaminants or specific contaminant groups. Canisters or cartridges are color coded to indicate the general chemicals or chemical class against which they are effective. Each sorbent can absorb only a certain amount of a contaminant, and after its saturation point is reached, contaminant "breakthrough" occurs; users in this case may be exposed to full atmospheric concentrations of the contaminant. Furthermore, chemical sorbent cartridges and canisters have an expiration date and once opened, humidity and other factors shorten their useful life. Canisters should therefore be used immediately and discarded after use.

Substances for which an air-purifying respirator is used should have adequate warning properties. According to NIOSH indications, this means the substance should have an odor, taste, or irritant effect that is detectable and persistent below the recommended exposure level. OSHA permits the use of air-purifying respirators in atmospheres contaminated with poor warning-property substances provided certain criteria are met, such as an approved end-of-service-life indicator (ESLI) on the respirator. Table 7.3 compares SCBAs, SARs, and air-purifying respirators.

7.6 REFERENCES

1. NIOSH, OSHA, U.S. Coast Guard, U.S. EPA. *Occupational Safety and Health Guidance Manual for Hazardous Waste Site Activities*. DHHS Publication No. 85-115, 1985.
2. U.S. EPA. *Hazardous Materials Incident Response Training Program*, 1983.

TABLE 7.3 Relative Advantages and Disadvantages of Respiratory Protective Equipment

Type of Respirator	Advantages	Disadvantages
Atmosphere-supplying SCBA	• Provides the highest available level of protection against airborne contaminants and oxygen deficiency. • Provides the highest available level of protection under strenuous work conditions.	• Bulky, heavy (up to 35 lb). • Finite air supply limits work duration. • May impair movement in confined spaces.
Positive-pressure SAR (also called air-line respirator)	• Enables longer work periods than SCBA. • Less bulky and heavy than SCBA. SAR equipment weighs less than 5 lb (or around 15 lb if escape SCBA protection is included). • Protects against most airborne contaminants.	• Not approved for use in IDLH atmosphere or in oxygen-deficient atmospheres unless equipped with emergency egress unit such as escape-only SCBA that can provide immediate emergency respiratory protection in case of air line failure. • Impairs mobility. • MSHA/NIOSH certification limits hose length to 300 ft (90 m). • As length of hose is increased, minimum approved air flow may not be delivered at facepiece. • Air line is vulnerable to damage, chemical contamination, and degradation. Decontamination of hoses may be difficult. • Worker must retrace steps to leave work area. • Requires supervision/monitoring of the air line.

Air-purifying		
Air-purifying respirator (including powered air-purifying respirators PAPRs])	• Enhanced mobility. • Lighter than a SCBA. Generally weighs 2 lb (1 kg) or less (except for PAPRs).	• Cannot be used in IDLH or oxygen-deficient atmospheres (<19.5% oxygen at sea level). • Limited duration of protection. May be hard to gauge safe operating time in field conditions. • Only protects against specific chemicals and up to specific concentrations. • Use requires monitoring of contaminant and oxygen levels. • Can be used only (1) against gas and vapor contaminants with adequate warning properties or (2) for specified gases and vapors provided that the service is known and a safety factor is applied or unit has an ESLI.

APPENDIX 7.1 SCBA CHECKOUT PROCEDURES

7.A.1 Introduction

Before a SCBA can be used, it must be properly inspected to help prevent malfunctions during use. The checklist that follows can help ensure proper inspection. The list is for pressure-demand SCBA units with no mode-select lever such as the MSA 401.

7.A.2 Checklist: Pressure-Demand SCBA without Mode Select Lever

Make sure that the following conditions exist before proceeding with the checklist:

- High-pressure hose connector is tight on cylinder fitting
- Bypass valve is closed
- Mainline valve is closed
- Regulator outlet is not covered or obstructed

A. Backpack and Harness Assembly
   1. Straps
      a. Visually inspect for complete set.
      b. Visually inspect for frayed or damaged straps.
   2. Buckles
      a. Visually inspect for mating ends.
      b. Check locking function.
   3. Back plate and cylinder lock
      a. Visually inspect back plate for cracks and missing rivets or screws.
      b. Visually inspect cylinder hold-down strap; physically check strap tightener and lock to ensure that it is fully engaged.

B. Cylinder and Cylinder Valve Assembly
   1. Cylinder
      a. Physically check to ensure that it is tightly fastened to a back plate.
      b. Visually inspect for large dents or gouges in metal.
      c. Check hydrostatic test date to ensure that it is current.
   2. Head and valve assembly
      a. Visually determine whether cylinder valve lock is present.
      b. Visually inspect cylinder gauge for condition of face, needle, and lens.

c. Open cylinder valve; listen or feel for leakage around packing. (If leakage is noted, do not use until repaired.) Note function of valve lock.

C. Regulator and High-Pressure Hose

1. High-pressure hose and connector

   Listen or feel for leakage in hose or at hose-to-cylinder connector. (A bubble in outer hose covering may be caused by seepage of air through hose when stored under pressure. This does not necessarily indicate a faulty hose.)

2. Regulator and low-pressure alarm

   a. Cover regulator outlet with palm of hand. Open mainline valve and read regulator gauge (must read at least 1800 psi and not more than rated cylinder pressure).
   b. Close cylinder valve and slowly move hand from regulator outlet to allow air to flow slowly. Gauge should begin to show immediate loss of pressure as air flows. Low-pressure alarm should sound between 650 and 550 psi. Remove hand completely from outlet and close mainline valve.
   c. Place mouth onto or over regulator outlet and blow. A positive pressure should be created and maintained for 5–10 sec without loss of air. Next, suck to create a slight negative pressure on regulator; hold for 5–10 sec. Vacuum should remain constant. This tests integrity of the diaphragm. Any loss of pressure or vacuum during this test indicates a leak in the apparatus.
   d. Open cylinder valve.
   e. Cover regulator outlet with palm of hand and open mainline valve. Remove hand from outlet and replace in rapid movement. Repeat twice. Air should escape when hand is removed each time, indicating a positive pressure in chamber. Close mainline valve and remove hand from outlet.
   f. Ascertain that the regulator outlet is not covered or obstructed. Open and close bypass valve momentarily to ensure flow of air through bypass system.

D. Facepiece and Corrugated Breathing Tube

1. Facepiece

   a. Visually inspect head harness for damaged serrations and deteriorated rubber. Visually inspect rubber facepiece body for signs of deterioration or extreme distortion.
   b. Visually inspect lens for proper seal in rubber facepiece, retaining clamp properly in place, and cracks or large scratches.
   c. Visually inspect exhalation valve for visible deterioration or buildup of foreign materials.

d. Perform negative pressure test for overall seal and check of exhalation valve. In monthly inspection, place mask against face and use the following procedure; in preparing for use, don backpack and then facepiece. With facepiece held tightly to face (or facepiece properly donned), stretch breathing tube to open corrugations and place thumb or hand over end of connector. Inhale. Negative pressure should be created inside mask, causing it to pull tightly to face for 5–10 sec. If negative pressure drops, do not wear facepiece.

2. Breathing tube and connector
   a. Stretch breathing tube and visually inspect for deterioration and holes.
   b. Visually inspect connector to ensure good condition of threads and for presence and proper condition of O-ring or rubber gasket seal.

E. Storage of Units

Certain criteria must be met before a SCBA is stored. Units not meeting the criteria should be set aside for repair by a certified technician.

1. Cylinder refilled as necessary and unit cleaned and inspected
2. Cylinder valve closed
3. High-pressure hose connector tight on cylinder
4. Pressure bled off of high-pressure hose and regulator
5. Bypass valve closed
6. Mainline valve closed
7. All straps completely loosened and laid straight
8. Facepiece properly stored to protect against dust, direct sunlight, extreme temperatures, excessive moisture, and damaging chemicals

8

Chemical Protective Clothing

8.1 INTRODUCTION

Chemical protective clothing prevents potentially hazardous chemicals from coming in contact with the human body. Because chemicals may be encountered in various forms and concentrations, predicting the exact hazards posed by released chemical agents is often difficult. It is thus important when making decisions about the kinds of chemical protective clothing to be worn that emergency response personnel and site workers:

- Thoroughly assess the nature and degree of the hazards present at an emergency
- Consider fully the tasks various response personnel will be expected to perform

8.2 SELECTION FACTORS

8.2.1 Chemical Resistance

The most important factor in selecting chemical protective clothing is the ability to withstand the extreme chemical effects encountered. It is of utmost importance, therefore, that the chemicals present at an emergency site be accurately identified as quickly as possible before personnel are committed to the site. Once chemicals at a site have been identified, standard references such as *Guidelines for the Selection of Chemical Protective Clothing*[1] can be consulted for detailed information regarding chemical resistance of various protective clothing materials.

The ability of protective clothing to prevent chemical contact with the body is a function of three key features: permeation, degradation, and penetration.

Permeation occurs when a contaminant contacts protective clothing material and establishes a concentration gradient. The contaminant is subsequently absorbed by the protective material and moves from the area of greater concentration (exterior surface) to the area of lesser concentration (interior surface). This movement is caused by diffusion and has been shown to be a unique feature of each different combination of protective material and chemical agent. Contact time and the contaminant concentration determine, to a great extent, the amount of permeation that can occur.

The degree of permeation ultimately determines the time required for a hazardous chemical to reach the interior surface—known as the breakthrough time. The permeation rate is determined by the chemical concentration (the more concentrated the hazardous chemical, the more likely that permeation will occur) and the thickness of the material (thicker material takes longer to break through). It is important to remember that no material is totally resistant to permeation by all hazardous chemicals; as described below, however, a number of protective clothing materials provide sufficient resistance for a wide range of substances likely to be encountered.

Once permeation has occurred, it may or may not be visually detectable. Because permeation is an essentially continuous process, proceeding until an equilibrium is established, harmful contaminant concentrations can develop on the inside surfaces in contact with the wearer. It should also be noted that decontamination does not immediately stop the permeation process. When a toxic agent is suspected of permeating protective clothing, the interior edge of clothing should also be checked for contamination before reuse, or the protective clothing should be immediately discarded.

A special note about chemical mixtures: Limited permeation data are available for mixtures of hazardous chemical agents. Thus, workers who may be exposed to chemical mixtures, such as hazardous wastes, should select protective clothing that offers the widest range of personal protection to chemicals suspected as mixture constituents. Also, protective clothing made from more than one material may offer a broader range of chemical protection.

Degradation results from a chemical reaction between protective clothing materials and hazardous chemical agents. These chemical reactions can produce the following:

- Shrinkage
- Swelling
- Softness
- Brittleness

These and other changes damage the integrity of the protective material and lead to greater permeation by chemical contaminants.

Penetration can result from the following:

- Imperfections in the design or construction features of protective clothing such as zippers and seams or in the material itself
- Rips, punctures, or tears

The integrity of protective clothing material must be checked prior to usage. Protective clothing from several different manufacturers should be examined to find the best quality control and design features for use at particular emergency response settings.

The criteria used for evaluations or rating systems on the chemical resistance of protective clothing materials often vary widely. Protective clothing material ratings are reported as fair, good, or excellent for a particular chemical agent. However, these ratings may be based on chemical degradation of clothing materials or physical changes in clothing materials and thus only provide an indication of breakthrough via permeation. Alternatively, evaluations of protective clothing may be based only on raw material and not on the actual clothing. Because this evaluation system does not consider material thickness, care should be used in applying such a rating system to all emergency response situations.

8.2.2 Durability

Response personnel and site workers are asked to perform a wide range of tasks while wearing protective clothing. Consideration must therefore be given to the ability of particular pieces of protective clothing to:

- Withstand the chemical hazards associated with tasks being carried out
- Resist puncture and tearing
- Withstand multiple contamination/decontamination procedures

8.2.3 Flexibility and Dexterity

Personnel wearing protective clothing must be able to reach, bend, and stretch with minimum difficulty. Consideration should also be given to determine if workers can use tools, instruments, and other equipment effectively, especially when gloves must be worn with the protective clothing.

8.2.4 Temperature Effects

Temperature extremes can change the ability of protective materials to resist permeation. Some synthetic polymers, for example, become very soft and flexible at high temperatures, which can affect the permeability of these materials. The increased pliability of synthetic materials at high temperatures can lead to stretching and eventually hinder movement. Alternatively, low temperatures cause some protective clothing materials to become rigid, thus hampering movement as well as the ease of donning equipment.

Many types of chemical protective clothing materials are virtually impermeable to moisture. Wearers often cannot be effectively cooled through the evaporation process while moving about during emergency operations. Therefore, lighter colors and lighter weights of material should be strongly considered for work in hot environments. It is also important to consider what undergarments are worn by emergency responders, especially when encapsulated suits are employed. Perspiration can be controlled by the worker wearing long cotton underwear, which can provide a small measure of cooling. Polymeric and elastic materials often become extremely hot in direct sunlight. Bare skin should be covered to prevent the possibility of thermal burns.

8.2.5 Capacity to Clean and Decontaminate Clothing

Currently, there are no universal design standards for chemical protective clothing. Care must be taken to ensure that a particular piece of clothing will meet all of the needs of a particular emergency—including the need to clean and decontaminate equipment. Studies should be conducted on site to ensure that cleaning and decontamination procedures do not pose chemical exposure hazards to emergency responders.

8.2.6 Compatibility with Other Equipment

It is also important to ensure that a protective clothing chosen for a particular situation will not preclude the use of other necessary equipment. For example, self-contained breathing apparatus (SCBA) units often have face-mounted regulators; encapsulated suits must be designed so that the facial area accommodates such regulators. Also, certain types of protective clothing may preclude the use of hard hats.

8.2.7 Shelf Life

Attention should be given to the shelf life of protective clothing and to the risk of deterioration through aging. Proper cleaning and equipment storage are important in ensuring the maximum useful life of protective clothing. Constant, long-term exposure to ultraviolet radiation (i.e., sunlight), mildew, and improper folding and hanging are some of the environmental factors that can easily damage protective clothing and render it useless. Many manufacturers now sell protective clothing with an expiration date to assist users in proper surveillance of the usable life of equipment.

8.3 COMPARISONS AND LEVELS OF PROTECTIVE EQUIPMENT

Table 8.1, adapted from *Guidelines for the Selection of Chemical Protective Clothing*,[1] provides a good comparison of the various types of chemical protective clothing materials and their physical characteristics.

8.4 SITE ENTRY—LEVELS OF PROTECTION

8.4.1 Introduction

This section describes the level of protection provided by protective clothing combinations or ensembles as developed by the EPA. Keep in mind that a particular ensemble must be chosen to fit a particular emergency site or situation. As the situation changes or as new information is obtained, it is often necessary to upgrade the level of personal protection clothing. Also, when possible, the level of protection should be downgraded to avoid undue stress on workers.

Personnel must wear protective equipment in the following hazardous response situations:

- Known or suspected atmospheric contamination
- Vapors, gases, or particulates
- When direct contact with toxic agents to the skin may occur

Equipment to protect the body against contact with known or anticipated chemical hazards has been divided into four categories according to the degree of protection afforded:

- **Level A**—Should be worn when the highest level of respiratory, skin, and eye protection is needed.
- **Level B**—Should be selected when the highest level of respiratory protection is needed, but a lesser level of skin protection is necessary. Level B protection is the minimum level recommended on initial site entries until the

TABLE 8.1 Physical Characteristics of Chemical Protective Clothing Materials[1]

Material	Abrasion Resistance	Cut Resistance	Flexibility	Heat Resistance	Ozone Resistance	Puncture Resistance	Tear Resistance	Relative Cost
Butyl rubber	F	G	G	E	E	G	G	High
Natural rubber	E	E	E	F	P	E	E	Medium
Neoprene	E	E	G	G	E	G	G	Medium
Neoprene/styrene butadiene rubber	G	G	G	G	G	G	G	Medium
Neoprene/natural rubber	E	E	E	G	G	G	G	Medium
Nitrile rubber	E	E	E	G	F	E	G	Medium
Nitrile rubber/ polyvinyl chloride	G	G	G	F	E	G	G	Medium
Polyethylene	F	F	G	F	F	P	F	Low
Chlorinated polyethylene	E	G	G	G	E	G	G	Low
Polyurethane	E	G	E	G	G	G	G	High
Polyvinyl alcohol	F	F	P	G	E	F	G	Very high
Polyvinyl chloride	G	P	F	P	E	G	G	Low
Styrene-butadiene rubber	E	G	G	G	F	F	F	Low
Viton	G	G	G	G	E	G	G	Very high

E, excellent; F, fair; G, good.

hazards have been further defined by on-site studies and appropriate personnel protection utilized.

- **Level C**—Should be selected when the type(s) of airborne substance(s) is known, the concentration(s) is measured, and the criteria for using air-purifying respirators are met.
- **Level D**—Should not be worn on any site with respiratory or skin hazards. It is primarily a work uniform providing minimal protection.

The level of protection selected should be based primarily on:

- Type(s) and measured concentrations(s) of the chemical substance(s) in the ambient atmosphere and its toxicity
- Potential or measured exposure to substances in air, splashes of liquids, or other direct contact with material due to work being performed

In situations where the type(s) of chemical(s), concentration(s), and possibilities of contact are not known, the appropriate level of protection must be selected based on professional experience and judgment until the hazards can be better characterized.

Although personal protective equipment reduces the potential for contact with harmful substances, ensuring the health and safety of response personnel requires, in addition, safe work practices, decontamination, site entry protocols, and other safety considerations. Together, these protocols establish a combined approach for reducing potential harm to workers.

8.4.2 Level A Protection

Items marked with an asterisk (*) are optional.

8.4.2.1 Personal Protective Equipment

- Pressure-demand SCBA approved by the Mine Safety and Health Administration (MSHA) and the National Institute for Occupational Safety and Health (NIOSH)
- Fully encapsulating vapor-tight, chemical-resistant suit
- Coveralls*
- Long cotton underwear*
- Gloves (outer), chemical resistant
- Gloves (inner), chemical resistant

- Boots, chemical resistant, steel toe and shank (depending on suit construction, worn over or under suit boot)
- Hard hat* (under suit)
- Disposable protective suit, gloves, and boots* (worn over fully encapsulating suit)
- Two-way radio

8.4.2.2 Criteria for Selection

Meeting any of the following criteria warrants use of Level A protection:

- The chemical substance(s) has been identified and requires the highest level of protection for skin, eyes, and the respiratory system based on:
  - Measured (or potential for) high concentration(s) of atmospheric vapors, gases, or particulates, or
  - Site operations and work functions involving high potential for splash, immersion, or exposure to unexpected vapors, gases, or particulates
- Extremely hazardous substances (e.g., dioxin, cyanide compounds, concentrated pesticides, DDT, poisonous gases, suspected carcinogens, and infectious substances) are known or suspected to be present, and skin contact is possible
- The potential exists for contact with substances that destroy skin
- Operations must be conducted in confined, poorly ventilated areas until the absence of hazards requiring Level A protection is demonstrated
- Total atmospheric readings indicate 500–1000 ppm of unidentified substances

8.4.2.3 Guidance on Selection Criteria

The fully encapsulating suit provides the highest degree of protection to the skin, eyes, and respiratory system if the suit material is resistant to the chemical(s) of concern during the time the suit is worn and/or at the measured or anticipated concentrations. Although Level A provides maximum protection, the suit material may be rapidly permeated and penetrated by certain chemicals from extremely high air concentrations, splashes, or immersion of boots or gloves in concentrated liquids or sludges. These limitations should be recognized when specifying the type of chemical-resistant garment. Whenever possible, the suit material should be matched with the substance it is used to protect against.

The use of Level A protection and other chemical-resistant clothing requires evaluating the problems of physical stress, in particular heat stress, associated with the wearing of impermeable protective clothing. Response personnel must be carefully monitored for physical tolerance and recovery. Because protective equipment is heavy and cumbersome, it decreases dexterity, agility, visual acuity, and so forth, and thus increases the probability of accidents. This probability decreases as less protective equipment is required. Thus, increased probability of accidents should be considered when selecting a level of protection. Many toxic substances are difficult to detect or measure in the field. When such substances (especially those readily absorbed by or destructive to the skin) are known or suspected to be present and personnel contact is unavoidable, Level A protection should be worn until more accurate information can be obtained.

8.4.3 Level B Protection

8.4.3.1 Personal Protective Equipment

- Pressure-demand SCBA (MSHA/NIOSH approved)
- Chemical-resistant clothing (overalls and long-sleeved jacket; coveralls; hooded, one- or two-piece chemical-splash suit; disposable chemical-resistant coveralls)
- Coveralls* (inner)
- Gloves (outer), chemical resistant
- Gloves (inner), chemical resistant
- Boots (outer), chemical resistant, steel toe and shank
- Boots (outer), chemical resistant (disposable)*
- Hard hat (face shield*)
- Two-way radio (intrinsically safe)

8.4.3.2 Criteria for Selection

Meeting any one of the following criteria warrants use of Level B protection:

- The type(s) and atmospheric concentration(s) of toxic substances have been identified and require the highest level of respiratory protection, but a lower level of skin and eye protection. These include atmospheres with any of the following characteristics:

- With immediately dangerous to life and health (IDLH) concentrations
- Exceeding limits of protection afforded by a full-face, air-purifying mask or containing substances for which air-purifying canisters do not exist
- Having low removal efficiency
- Containing substances requiring air-supplied equipment, but substances and/or concentrations do not represent a serious skin hazard

- The atmosphere contains less than 19.5% oxygen
- Site operations make it highly unlikely that the small, unprotected area of the head or neck will be contacted by splashes of extremely hazardous substances
- Total atmospheric concentrations of unidentified vapors or gases range from 5 to 500 ppm, and vapors are not suspected of containing high levels of chemical toxic to skin

8.4.3.3 Guidance on Selection Criteria

Level B equipment provides a high level of protection to the respiratory tract but a somewhat lower level of protection to skin. The chemical-resistant clothing required in Level B is available in a wide variety of styles, materials, construction detail, permeability, etc. Each of these factors affects the degree of protection afforded. Therefore, a specialist should select the most effective chemical-resistant clothing (and fully encapsulating suit) based on the known or anticipated hazards and/or job function.

Level B skin protection is selected by:

- Comparing the concentrations of known or identified substances in air with skin toxicity data
- Determining the presence of substances that are destructive to and/or readily absorbed through the skin by liquid splashes, unexpected high levels of gases or particulates, or other means of direct contact
- Assessing the effect of the substance (at its measured air concentrations or splash potential) on the small area of the head and neck unprotected by chemical-resistant clothing.

For initial site entry and reconnaissance at an open site, approaching whenever possible from the upwind direction, Level B protection (with good quality, hooded, chemical-resistant clothing) should protect response personnel, provided the conditions described in selecting Level A are known or judged to be absent. For continuous operations, the aforementioned criteria must be evaluated. At 500 ppm total vapors/gases, upgrading to Level A protection may be advisable. A major factor for re-evaluation is the presence of vapors, gases, or particulates requiring a higher degree of skin protection.

8.4.4 Level C Protection

8.4.4.1 Personal Protective Equipment

- Full-face, air-purifying, canister-equipped respirator (MSHA/NIOSH approved)
- Chemical-resistant clothing (coveralls; hooded, two-piece chemical splash suit; chemical-resistant hood and apron; disposable chemical-resistant coveralls)
- Coveralls* (inner)
- Gloves (outer), chemical resistant
- Gloves (inner), chemical resistant*
- Boots (outer), chemical resistant, steel toe and shank*
- Boots (outer), chemical resistant (disposable)*
- Hard hat (face shield*)
- Two-way radio

8.4.4.2 Criteria for Selection

Meeting all of the following criteria permits use of Level C protection:

- Measured concentrations of identified substances will be reduced by the respirator to at or below the substance's exposure limit, and the concentration is within the service limit of the canister
- Atmospheric contaminant concentrations do not exceed IDLH levels
- Atmospheric contaminants, liquid splashes, or other direct contact will not adversely affect the small area of skin left unprotected by chemical-resistant clothing
- Job functions have been determined not to require SCBAs
- Total vapor readings register between background and 5 ppm above background
- Air will be monitored periodically

8.4.4.3 Guidance on Selection Criteria

Level C protection is distinguished from Level B by the equipment used to protect the respiratory system, assuming the same type of chemical-resistant clothing is used. The main selection criterion for Level C is that conditions permit wearing air-purifying devices.

The air-purifying device must be a full-face mask (MSHA/NIOSH approved) equipped with a canister suspended from the chin or on a harness. Canisters must be able to remove the substances encountered. Quarter or half masks or cheek-cartridge full-face masks should be used only with the approval of a qualified individual.

In addition, a full-face, air-purifying mask can be used only if:

- The oxygen content of the atmosphere is at least 19.5% by volume. Substance(s) is identified and its concentration(s) measured. Substance(s) has adequate warning properties.
- The individual passes a qualitative fit test for the mask.
- An appropriate cartridge/canister is used and its service limit concentration is not exceeded.
- An air monitoring program is part of all response operations when atmospheric contamination is known or suspected. It is particularly important that the air be monitored thoroughly when personnel are wearing air-purifying respirators (Level C). Continual surveillance using direct-reading instruments and air sampling is needed to detect any protection.
- Total unidentified vapor/gas concentrations of 5 ppm above background require Level B protection. Only a qualified individual should select Level C (air-purifying respirators) protection for continual use in an unidentified vapor/gas concentration of background to 5 ppm above background.

8.4.5 Level D Protection

8.4.5.1 Personal Protective Equipment

- Coveralls
- Gloves*
- Boots/shoes, leather or chemical resistant, steel toe and shank
- Boots (outer), chemical resistant (disposable)*
- Safety glasses or chemical splash goggles*
- Hard hat (face shield*)
- Escape mask*

8.4.5.2 Criteria for Selection

Meeting either of the following criteria allows use of Level D protection:

- No hazardous chemicals or pollutants have been measured
- Work functions preclude splashes, immersion, or potential for unexpected inhalation of any chemicals

8.4.5.3 Guidance on Selection Criteria

Level D protection is primarily a work uniform. It can be worn in areas where:

- Only boots can be contaminated
- There are no inhalable toxic substances

8.5 REFERENCES

1. A. D. Swope, P. P. Costas, J. O. Jackson, and D. J. Weitsman. *Guidelines for the Selection of Chemical Protective Clothing, Volume I: Field Guide*. Little, Brown, Cambridge, MA, 1983.
2. U.S. EPA. *Hazardous Materials Incident Response Training Program*, 1983.
3. J. V. Friel, M. J. McGoff, and S. J. Rodgers. *Material Development Study for a Hazardous Chemical Protective Clothing Outfit*. MSA Research Corp., Evans City, PA, 1980.
4. E.I. DuPont de Nemours. *Response to Transportation Emergencies, Unit 3, Book 2, Rhythm Individualized Training*. Wilmington, DE.

Chemical Decontamination: Principles and Practices

9.1 INTRODUCTION

The process of decontamination consists of physically removing or neutralizing hazardous contaminants from personnel and equipment to prevent continued contact of harmful substances with protective clothing, protective breathing apparatus, and other equipment. One additional rationale for proper decontamination is to minimize contaminant transfer from controlled areas to response personnel or site workers.

Decontamination procedures vary depending on the hazardous contaminants encountered at particular worksites:

- As expected, highly toxic substances, in general, require more detailed and thorough "decon" methods than chemicals that are minimally toxic.
- The particular steps, equipment, materials, and procedures employed must also be judged separately for each site or incident.

With an understanding of the general principles and guidelines involved in decontamination, specific procedures can be implemented according to particular situations.

9.2 PLANNING FOR DECONTAMINATION ACTIVITIES

One of the most important parts of the standard operating safety procedures at any incident or hazardous waste site are the methods to minimize contamination. A number of routine procedures can greatly reduce the amount and extent of personnel decontamination that is ultimately required:

- Use of remote sampling devices
- Properly bagging monitoring instruments
- Ensuring that personnel do not walk through areas of obvious contamination
- Requiring responders not to touch contaminated surfaces unless necessary and to wear disposable outer garments

Before any personnel are allowed access into areas where contamination might occur, a decontamination plan and its necessary procedures should be in place. Items commonly considered in developing a plan include:

- What are the most appropriate decontamination methods?
- What equipment will be needed?
- How many stations will be used?
- How will the decontamination area be laid out?
- What methods can be used to control contamination in the decon area?
- How will cleaning solutions, clothing, and other partially decontaminated items be handled?

The area designated for decontamination is usually referred to as the contamination reduction corridor (see Figure 9.1). Access to the area should be provided along a straight path and limited to authorized personnel only. The size of the contamination reduction corridor often varies with the number of decontamination stations needed, the number of workers requiring decon, and the amount of space available. An area 75 × 15 ft should accommodate full decontamination gear, and, ideally, a space of 3 ft should separate each decontamination station.

The contamination reduction corridor should be clearly marked so that entry and exit points can be carefully monitored. All personnel leaving the work zone should exit through the contamination reduction corridor; active personnel in the contamination reduction corridor must wear all required PPE. An exclusive area for segregating tools, contaminated disposable clothing, and other items must be established to prevent equipment cross-contamination; an additional corridor may be designated for heavy equipment.

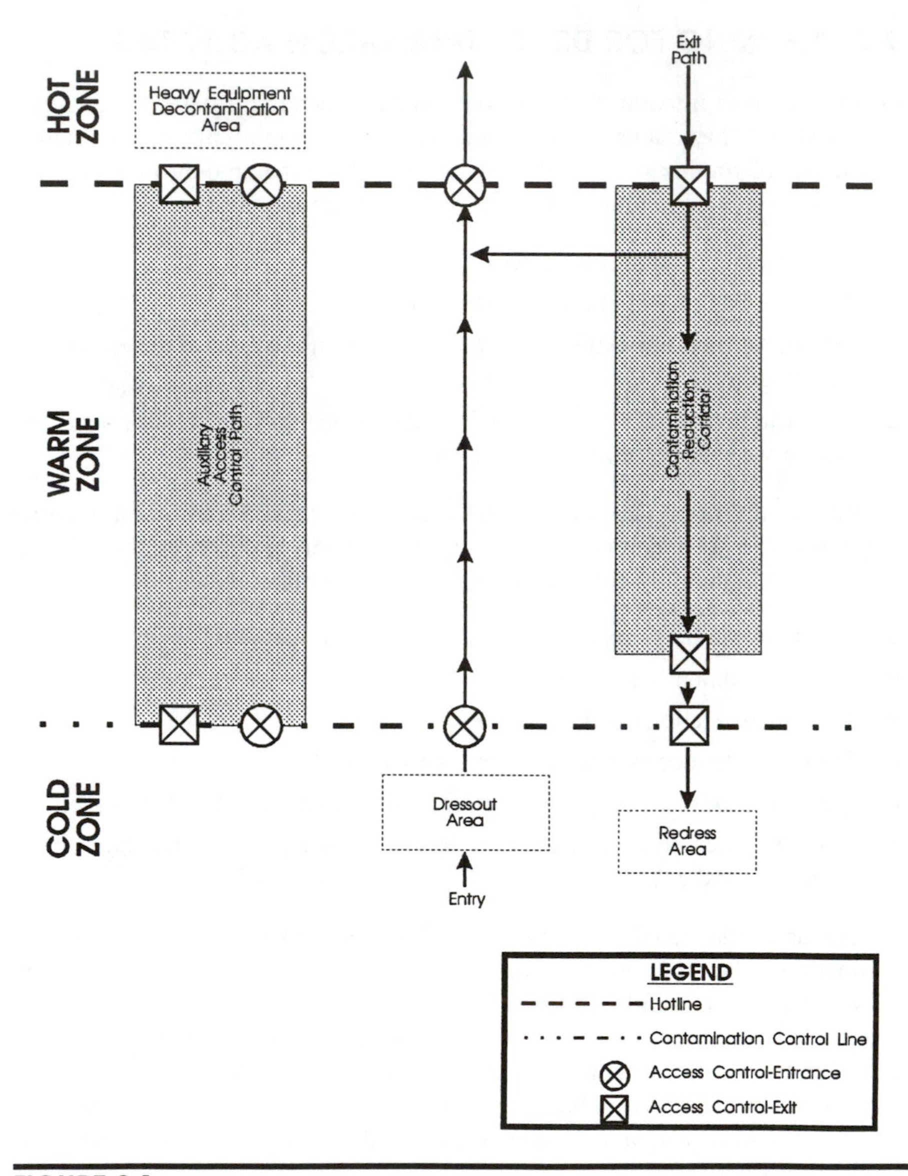

FIGURE 9.1 Site Work Zones

A number of factors may suggest modifications in the general decontamination plan, including:

- The type of contaminant found in the area
- The type of PPE worn

When a worker temporarily exits the exclusion zone, decontamination (or decon) does not need to be as thorough as for workers leaving the site at the end of a work shift. Ideally, different decontamination personnel should be located at each station to avoid direct contact with contaminants found at particular decontamination sites; however, this is not always possible for practical reasons. When deciding on the final design of the reduction contamination corridor, one should weigh the benefits of saving time and space against the consequences of reducing worker safety.

9.3 DECONTAMINATION PROCEDURES

The objective of decontamination is to sequentially reduce the level of contamination of all personnel, equipment, and other materials leaving the contaminated area. Physical decontamination may be accomplished by brushing, wiping, rinsing (high pressure or gravity), and/or heating (steam cleaning equipment only). If physical methods are not effective, wash/rinse procedures using chemical cleaning solutions should be considered. However, chemical solvents, neutralization solutions, and other reactive agents can be hazardous to workers and damaging to PPE. Figure 9.2 describes the steps in the decontamination process for levels of protection A to C, including modifications which may be required in particular cases with specific hazardous agents.

Workers performing decontamination procedures must also be protected from possible exposure to hazardous substances. The level of protection necessary will vary depending on several factors:

- Type of contaminants handled
- Expected degree of contamination
- Decon procedures employed

Personnel in the final decon stages, for example, require less protection than those involved in the initial steps.

9.4 DECONTAMINATION EFFECTIVENESS

Visual inspection of protective clothing will often detect discoloration, evidence of material degradation, or even contaminants still adhering to the clothing. Because small amounts of highly toxic agents may not be visible, however, observation by itself is not an acceptable method to determine decontamination effectiveness and efficiency. An alternative method is swipe testing. A cloth, piece of glass fiber filter paper, or swab is applied over a

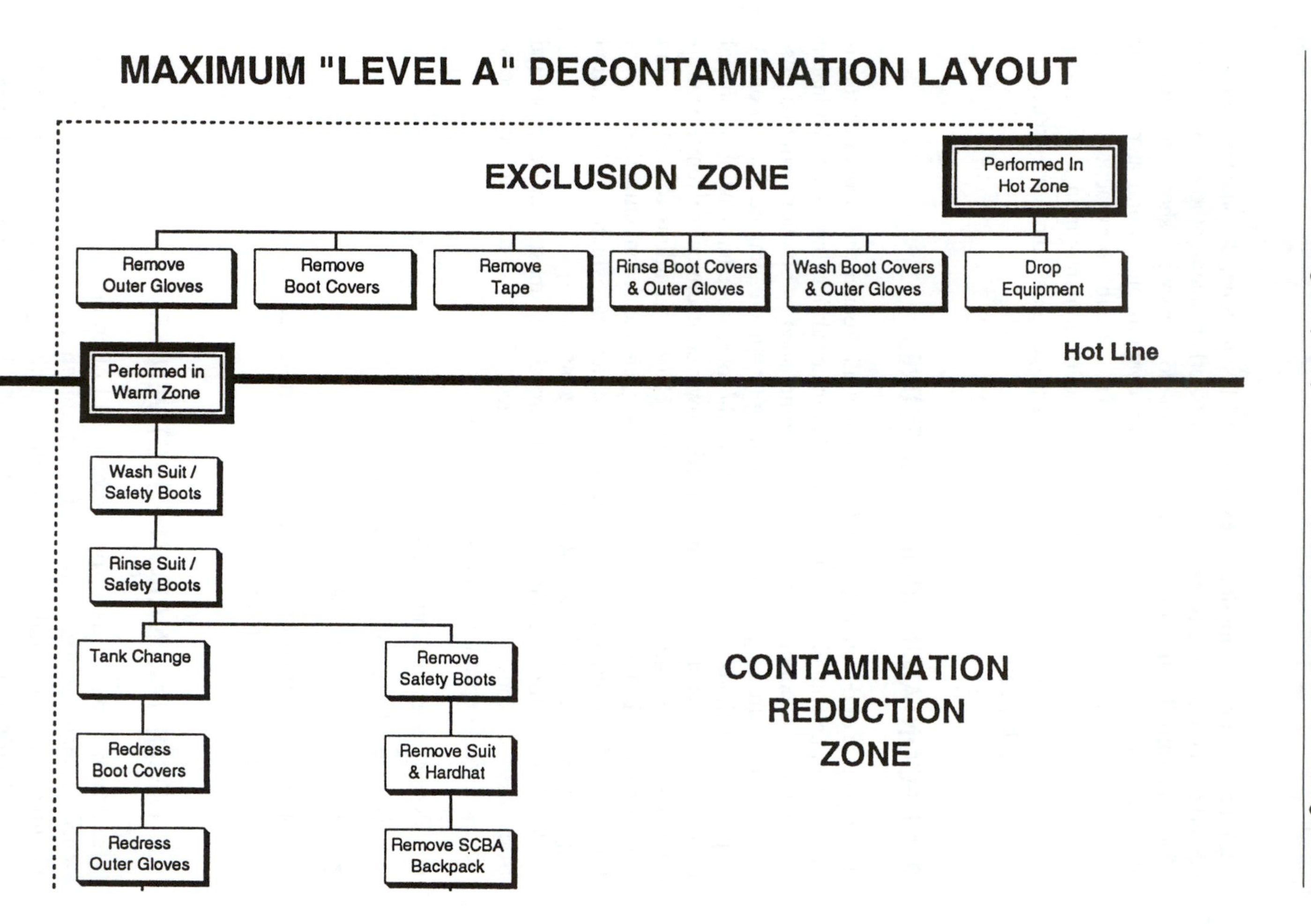
MAXIMUM "LEVEL A" DECONTAMINATION LAYOUT
EXCLUSION ZONE
Performed In Hot Zone
Remove Outer Gloves
Remove Boot Covers
Remove Tape
Rinse Boot Covers & Outer Gloves
Wash Boot Covers & Outer Gloves
Drop Equipment
Hot Line
Performed in Warm Zone
Wash Suit / Safety Boots
Rinse Suit / Safety Boots
Tank Change
Redress Boot Covers
Redress Outer Gloves
Remove Safety Boots
Remove Suit & Hardhat
Remove SCBA Backpack
CONTAMINATION REDUCTION ZONE

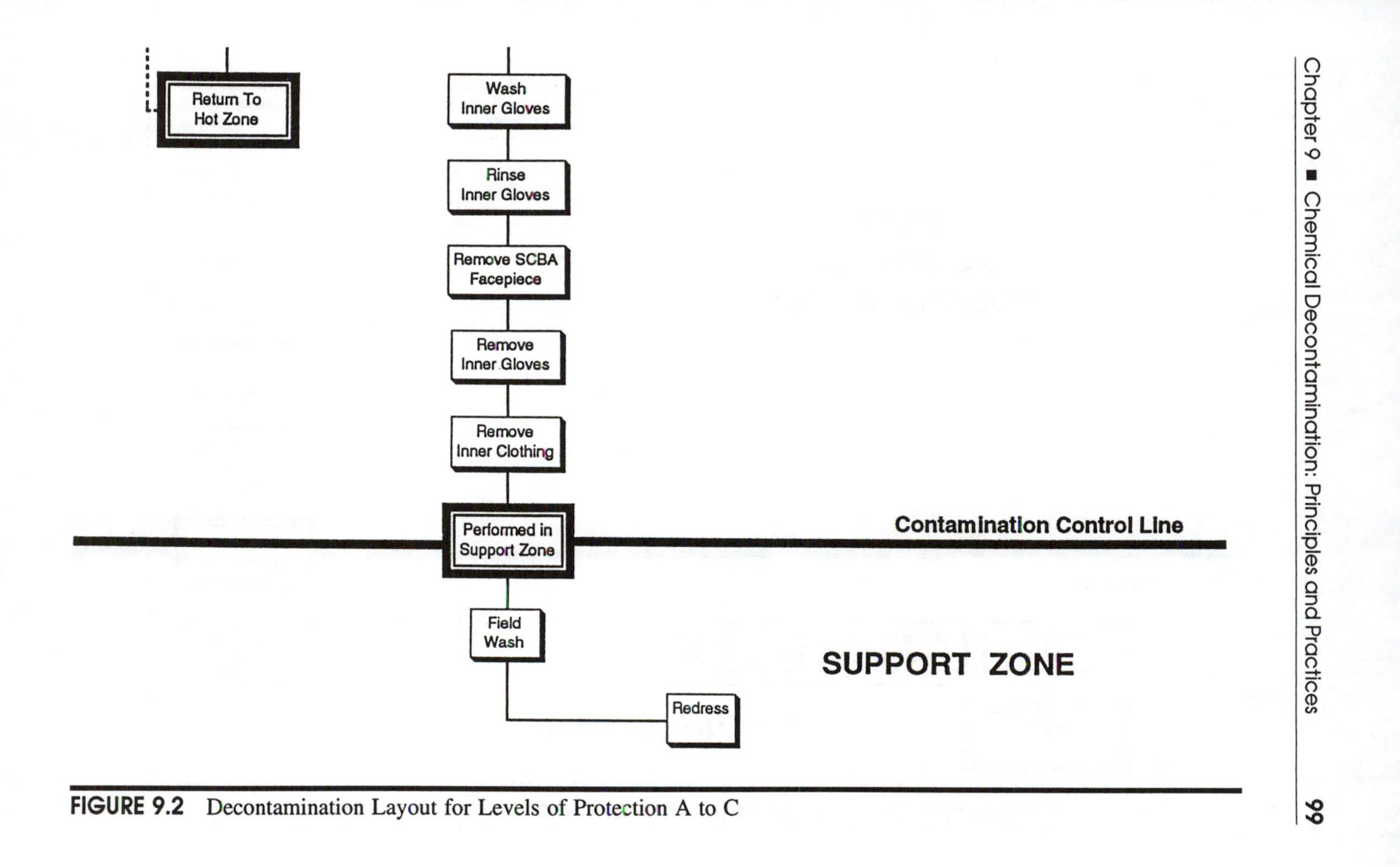

FIGURE 9.2 Decontamination Layout for Levels of Protection A to C

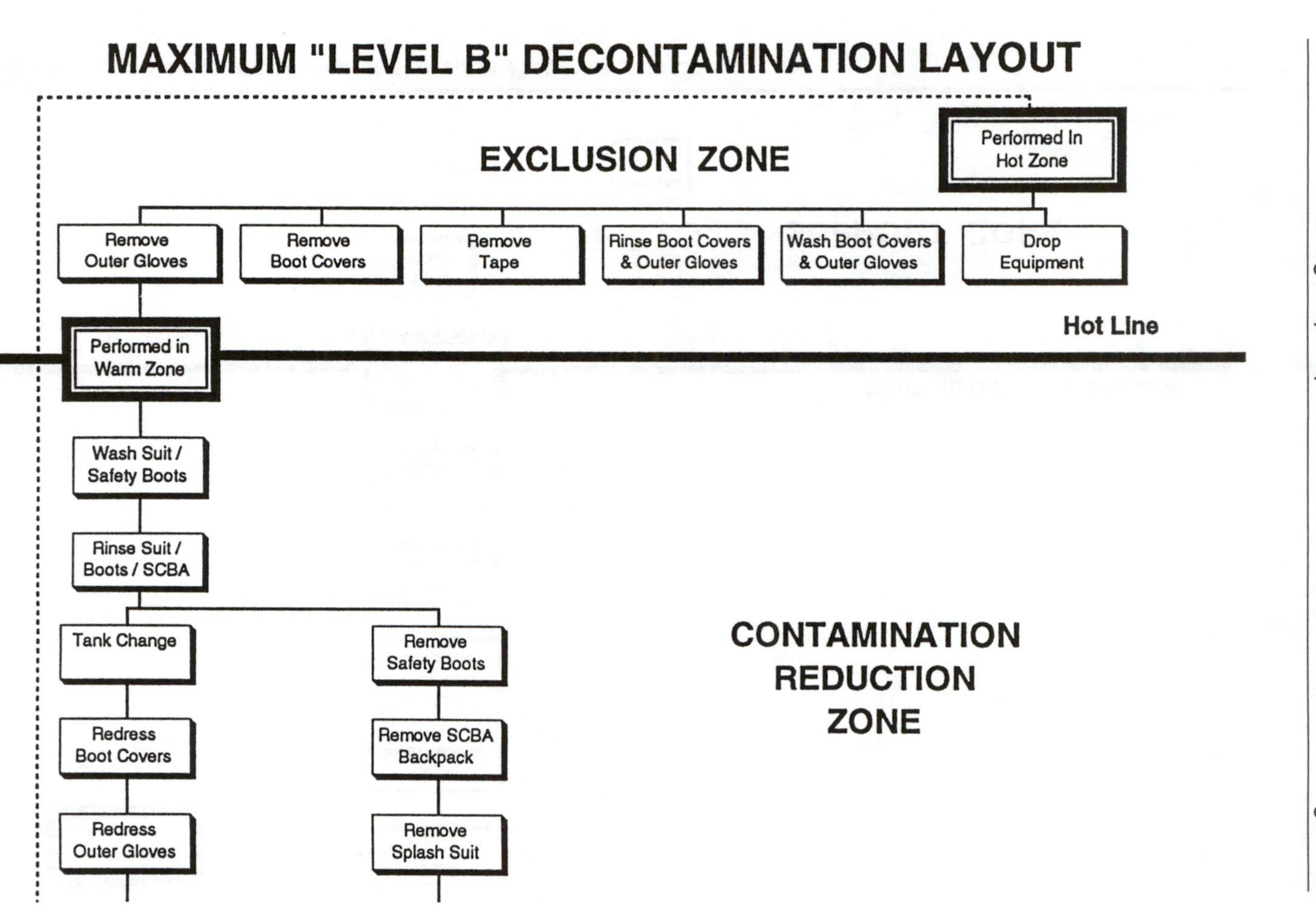

MAXIMUM "LEVEL B" DECONTAMINATION LAYOUT
EXCLUSION ZONE
Performed In Hot Zone
Remove Outer Gloves
Remove Boot Covers
Remove Tape
Rinse Boot Covers & Outer Gloves
Wash Boot Covers & Outer Gloves
Drop Equipment
Hot Line
Performed in Warm Zone
Wash Suit / Safety Boots
Rinse Suit / Boots / SCBA
Tank Change
Redress Boot Covers
Redress Outer Gloves
Remove Safety Boots
Remove SCBA Backpack
Remove Splash Suit
CONTAMINATION REDUCTION ZONE

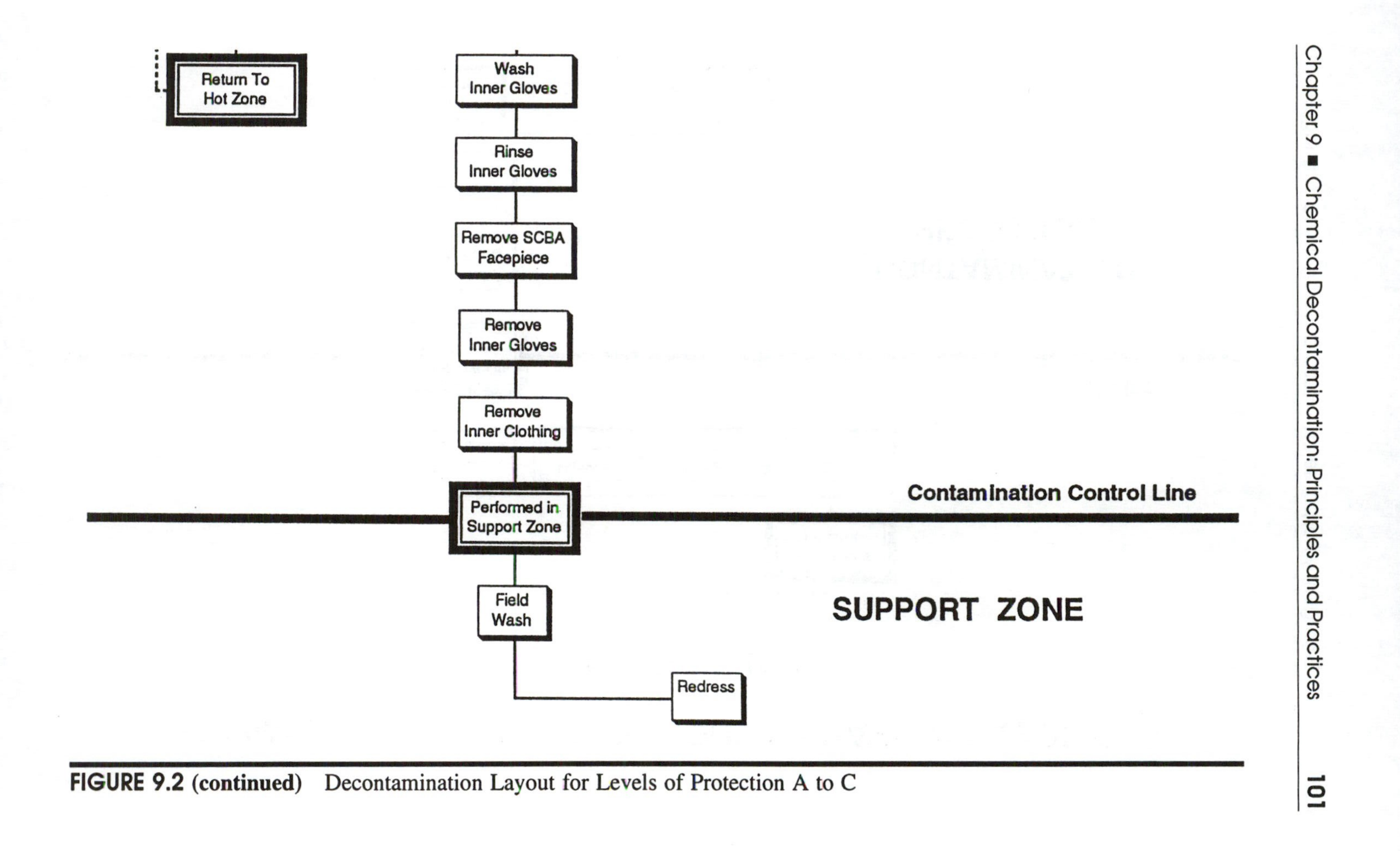

FIGURE 9.2 (continued) Decontamination Layout for Levels of Protection A to C

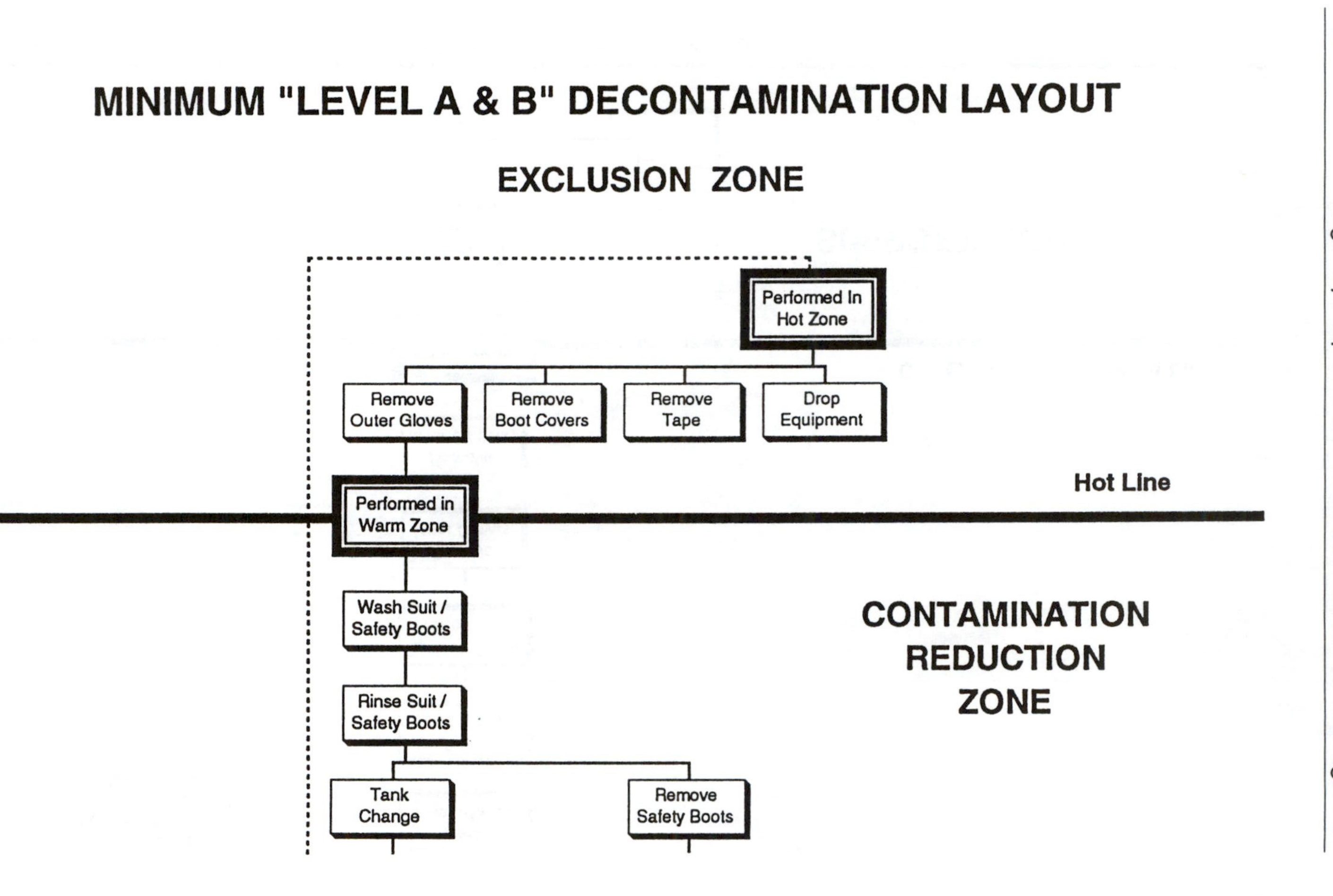
MINIMUM "LEVEL A & B" DECONTAMINATION LAYOUT
EXCLUSION ZONE
Performed In Hot Zone
Remove Outer Gloves
Remove Boot Covers
Remove Tape
Drop Equipment
Hot Line
Performed in Warm Zone
Wash Suit / Safety Boots
Rinse Suit / Safety Boots
Tank Change
Remove Safety Boots
CONTAMINATION REDUCTION ZONE

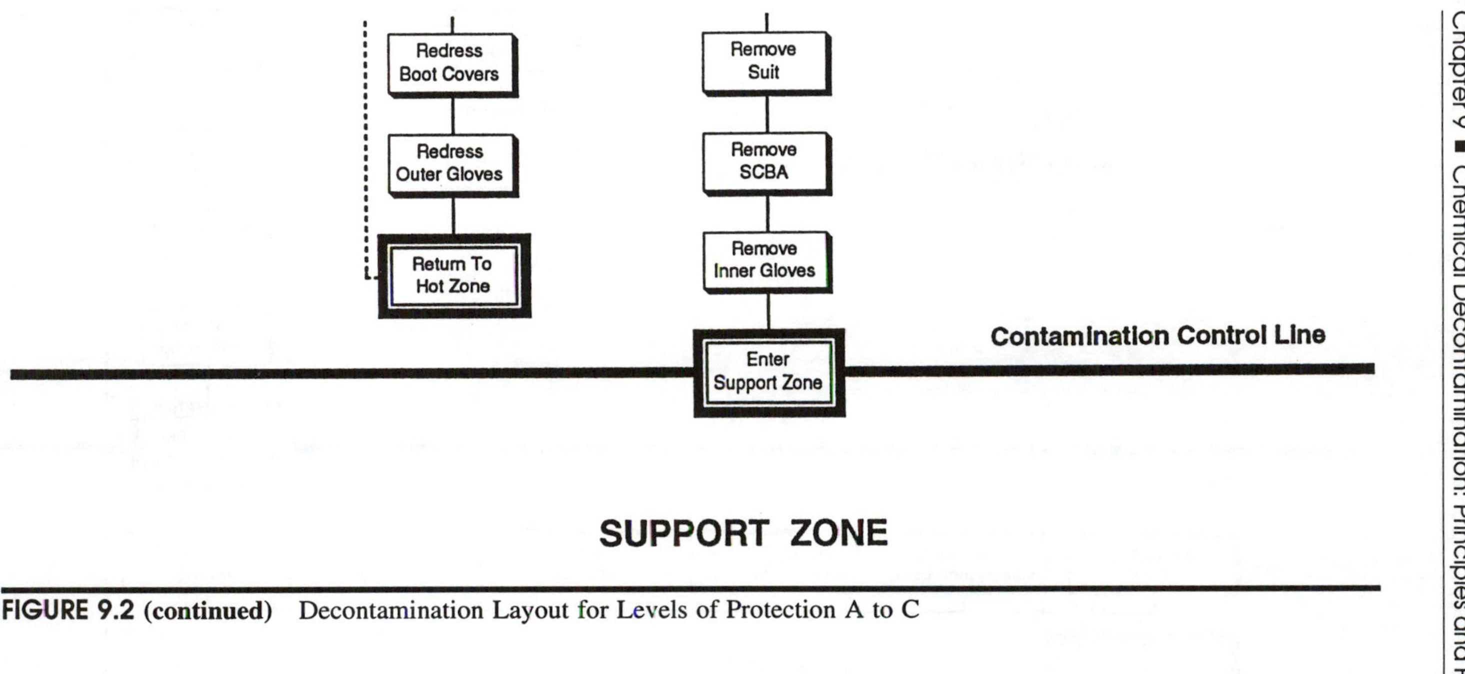

FIGURE 9.2 (continued) Decontamination Layout for Levels of Protection A to C

MAXIMUM "LEVEL C" DECONTAMINATION LAYOUT

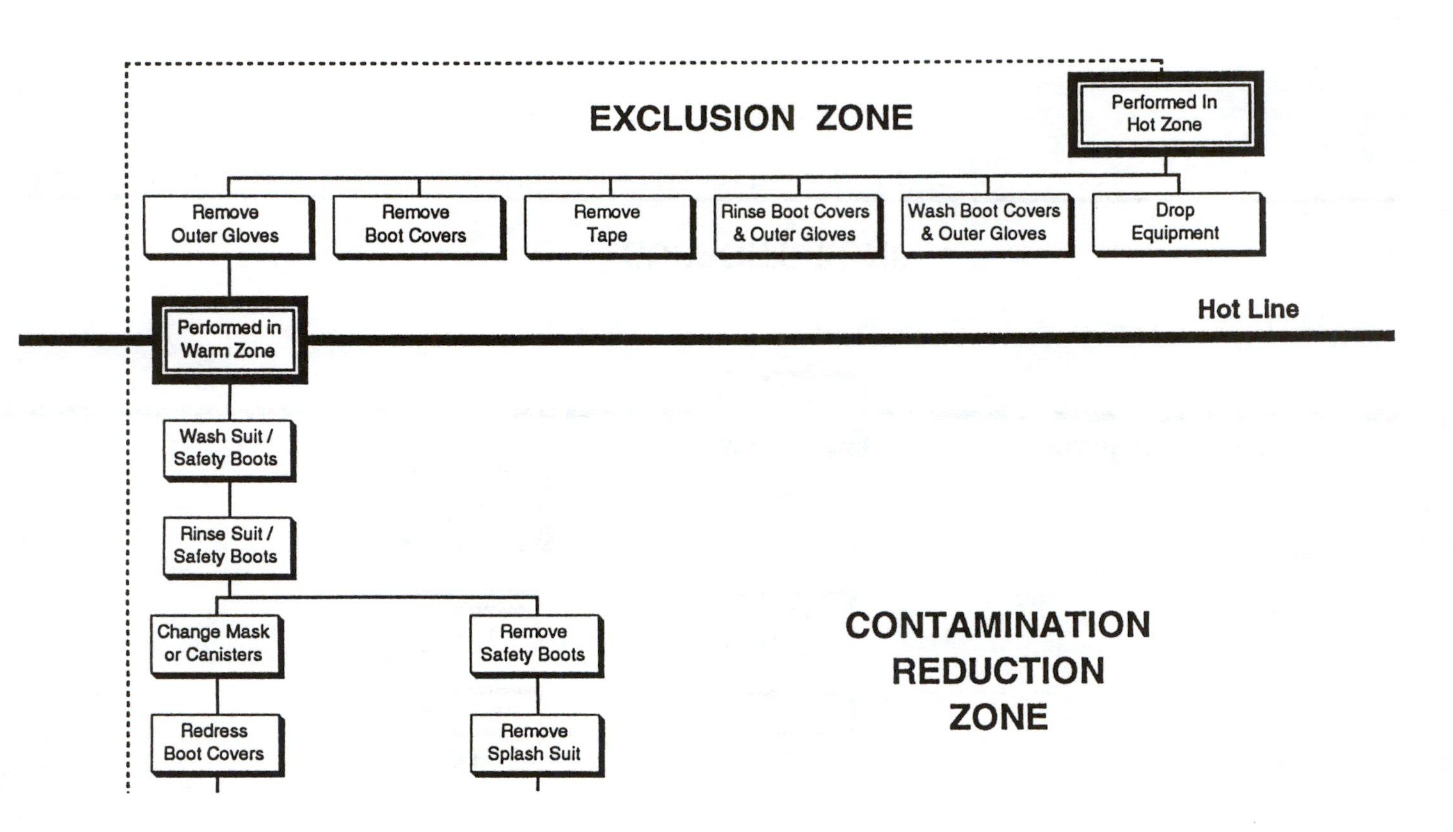
EXCLUSION ZONE
Performed In Hot Zone
Remove Outer Gloves
Remove Boot Covers
Remove Tape
Rinse Boot Covers & Outer Gloves
Wash Boot Covers & Outer Gloves
Drop Equipment
Hot Line
Performed in Warm Zone
Wash Suit / Safety Boots
Rinse Suit / Safety Boots
Change Mask or Canisters
Remove Safety Boots
Redress Boot Covers
Remove Splash Suit
CONTAMINATION REDUCTION ZONE

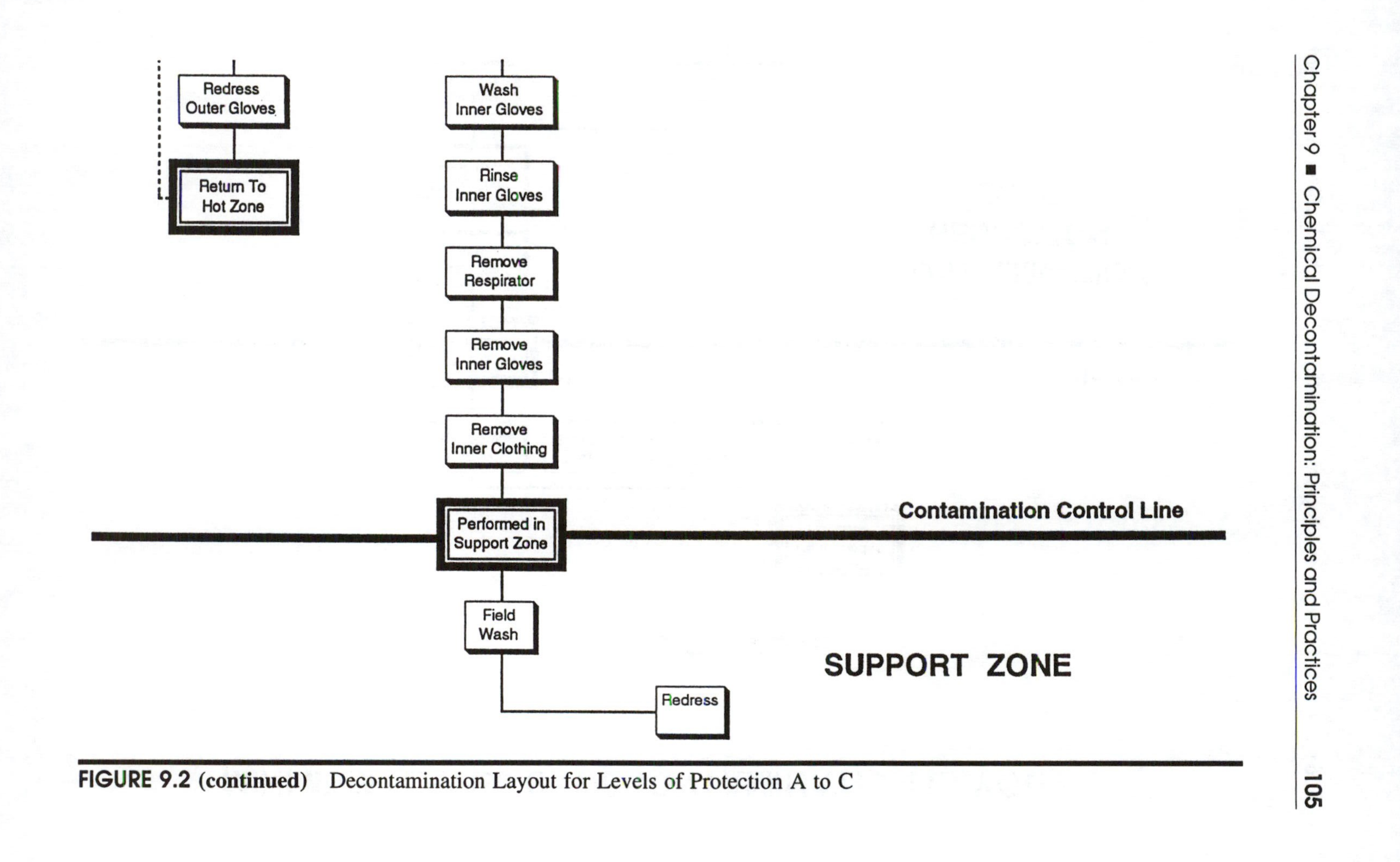

FIGURE 9.2 (continued) Decontamination Layout for Levels of Protection A to C

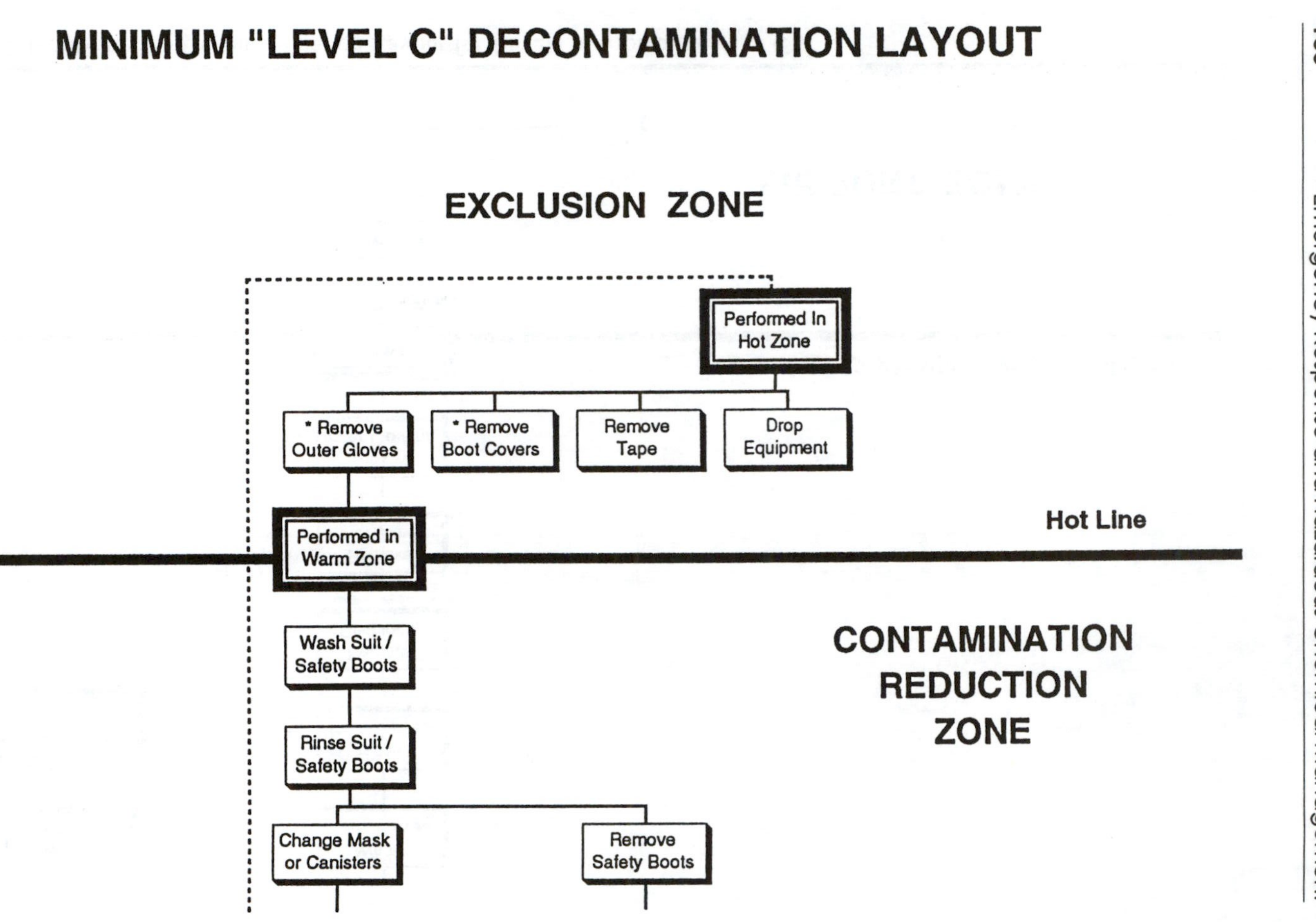

MINIMUM "LEVEL C" DECONTAMINATION LAYOUT
EXCLUSION ZONE
Performed In Hot Zone
* Remove Outer Gloves
* Remove Boot Covers
Remove Tape
Drop Equipment
Hot Line
Performed in Warm Zone
Wash Suit / Safety Boots
Rinse Suit / Safety Boots
Change Mask or Canisters
Remove Safety Boots
CONTAMINATION REDUCTION ZONE

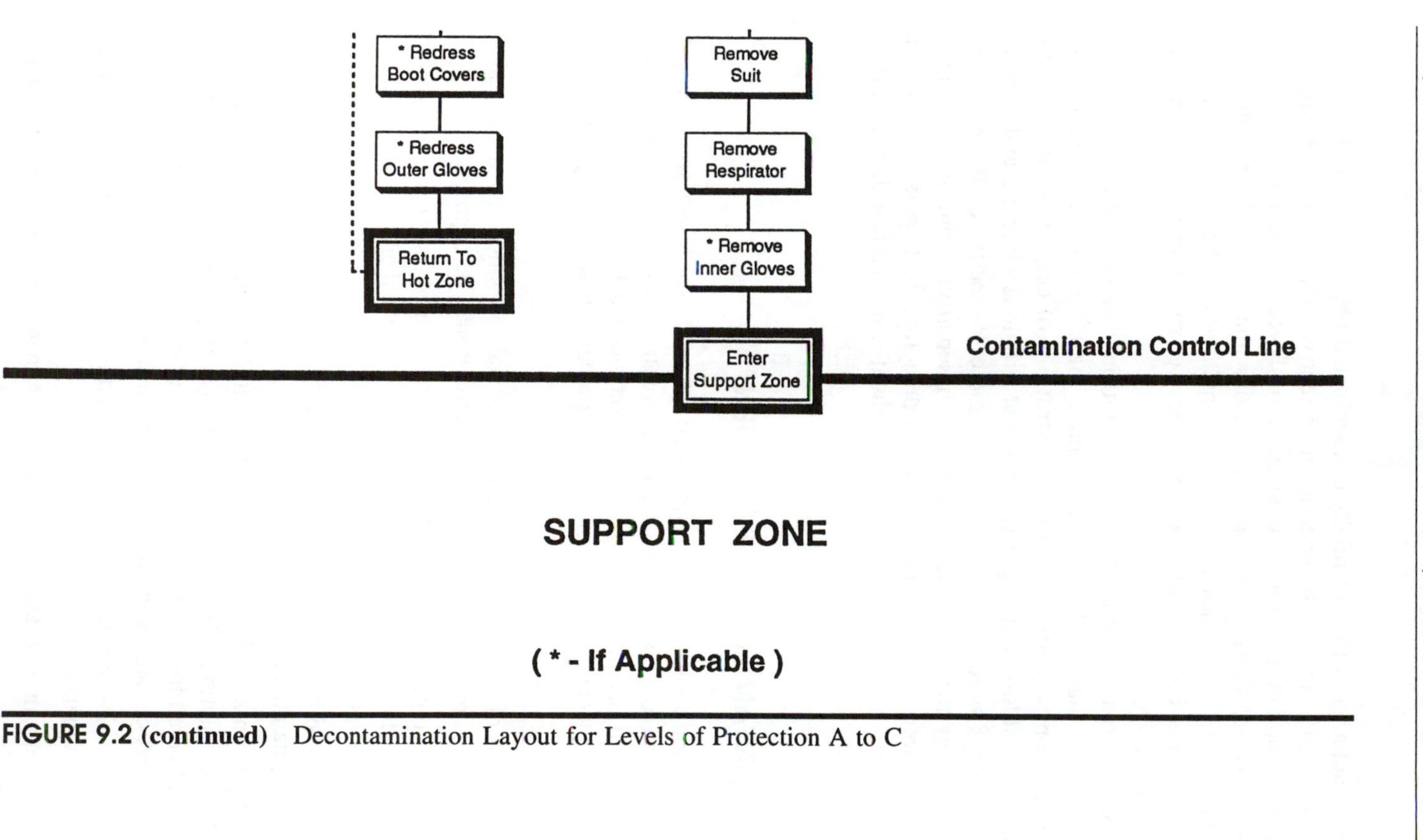

FIGURE 9.2 (continued) Decontamination Layout for Levels of Protection A to C

predetermined area of the potentially contaminated object to ascertain if contaminants are still present. Swipe tests on protective clothing may be done on both the inner and outer surfaces or on the skin of workers who have removed their protective clothing. Subsequent laboratory analysis of these swipes or swabs is necessary to evaluate decontamination; however, the time required for laboratory analysis often limits the applicability of swipe testing to the evaluation of alternative decon methods.

Laboratory analysis of the final rinse solution used in decontamination is another method that can be used to evaluate the effectiveness of particular decontamination procedures. High concentrations of contaminants found in the final rinse solution strongly suggest that previous cleaning and decontamination steps were not successful or complete. Protective clothing, equipment, and other tools may be surveyed with sensitive monitoring devices such as photoionization or flame ionization detectors. Hot spots or detection of unacceptably high contaminant levels indicates inadequate decontamination procedures.

9.5 DECONTAMINATION EQUIPMENT

Decontamination equipment and materials do not need to be highly sophisticated or expensive, although specially designed decon trailers and other systems are now commercially available. The following recommended decontamination equipment for personnel and for PPE can be easily procured in almost any area of the country:

- Drop cloths of plastic or other suitable materials on which heavily contaminated equipment and outer protective clothing may be deposited
- Collection containers, such as drums or suitable lined trash cans, for storing disposable clothing and heavily contaminated personal protective clothing or equipment that must be discarded
- Lined box with absorbents for wiping or rinsing off gross contaminants and liquid contaminants
- Large galvanized tubs, stock tanks, or children's wading pools to hold wash and rinse solutions. These should be at least large enough for a worker to place a booted foot in and should have either no drain or a drain connected to a collection tank or appropriate water treatment system.
- Wash solutions selected to wash off and reduce the hazards associated with the contaminants
- Rinse solutions selected to remove contaminants and contaminated wash solutions

- Long-handled, soft-bristled brushes to help wash and rinse off contaminants
- Paper or cloth towels for drying protective clothing and equipment
- Lockers and cabinets for storage of decontaminated clothing and equipment
- Metal or plastic cans or drums for contaminated wash and rinse solutions
- Plastic sheeting, sealed pads with drains, or other appropriate methods for containing and collecting contaminated wash and rinse solutions spilled during decontamination
- Shower facilities for full body wash or, at minimum, personal wash sinks (with drains connected to a collection tank or appropriate water treatment system)
- Soap or wash solution, wash cloths, and towels for personnel
- Lockers or closets for clean clothing and personal item storage

9.6 DECONTAMINATION SOLUTIONS

Decontamination solutions typically consist of a detergent/water mixture that is administered with a soft-bristle brush and removed by rinsing with copious amounts of clean water. Although this method is safer than using chemical solvents, certain contaminants may be water reactive and others may not be cleaned effectively by this method. Whenever decontamination with detergents and water proves ineffective, an experienced chemist should be consulted to suggest alternative chemical methods. Decontamination solutions for various classes of hazardous substances are listed in Table 9.1. Guidelines for preparing decontamination solutions are provided in Section 9.9.

9.7 EMERGENCY DECONTAMINATION

Medical emergencies must be handled according to special procedures that require the following issues be addressed:

- Decontamination should take place in a timely manner and without increasing the victim's risk of permanent damage.
- Attending medical personnel should be well protected.
- Contaminated protective and wash solutions should be disposed of properly.

Immediate decontamination is not the best course of action if it aggravates an accident victim's injuries or delays life-saving treatment. When possible, however, decontamination should be performed as soon as possible. Figure 9.3

TABLE 9.1 General Solubility of Contaminants in Four Solvent Types

Solvent	Soluble Contaminants
Water	Short-chain hydrocarbons Inorganic compounds Salts Some inorganic acids and polar agents
Dilute acids	Caustic compounds Amines Hydrazines
Dilute bases (detergents and soaps)	Acidic compounds Phenols Thiols Some nitro and sulfonic compounds
Organic solvents* Alcohols Ethers Ketones Aromatics Straight-chain aliphatics (hexane) Common petroleum products (fuel oil, kerosene)	Nonpolar organic compounds

* *Warning*: Some organic solvents can permeate and/or degrade the protective clothing.

shows a decision diagram for emergency decontamination situations. When decontamination cannot be done because of interference with emergency treatment, the victim should be wrapped in plastic, blankets, or other materials to reduce the spread of the contamination to other personnel. Medical personnel should be informed about specific decontamination procedures practiced on victims prior to their arrival and warned about the potential for chemical contamination.

9.8 DECONTAMINATION OF RESPONSE EQUIPMENT

All equipment taken into the exclusion zone (area where hazardous chemicals are handled) and the chemical reduction corridor is subject to contamination. Delicate equipment and instruments should thus be protected whenever possible by employing plastic bags to cover openings and vents. Because tools with wooden

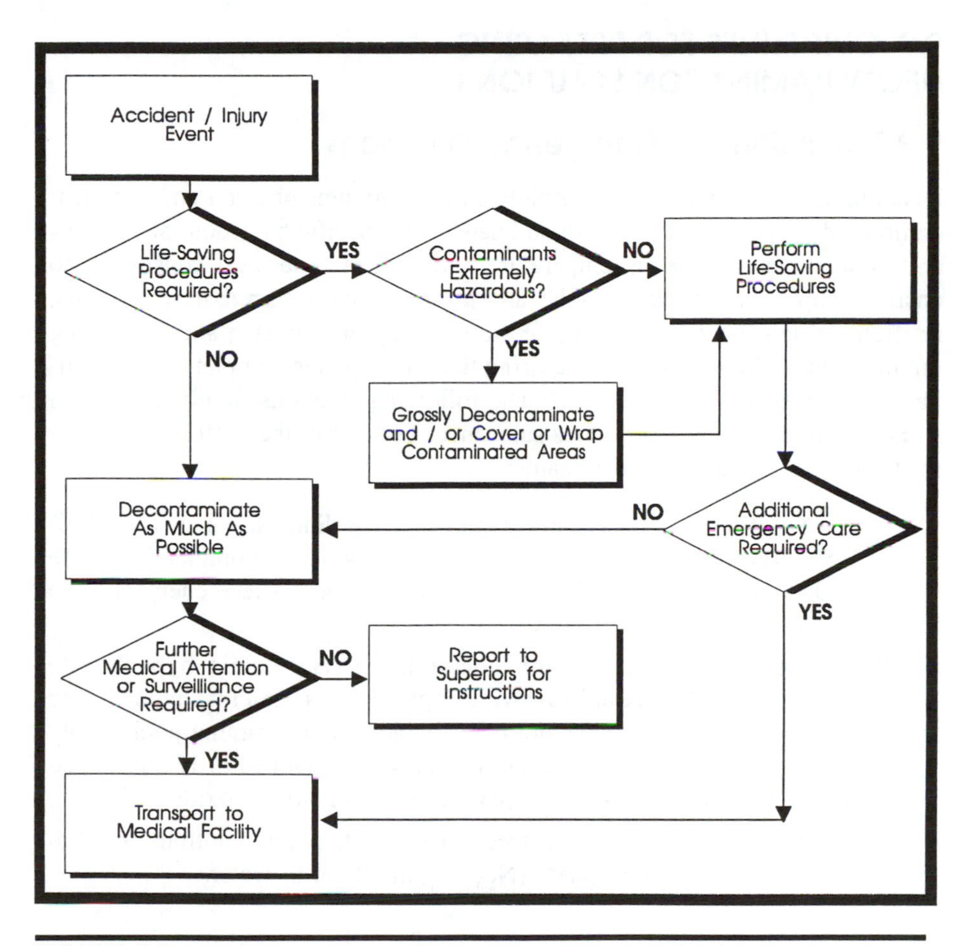

FIGURE 9.3 Decision Diagram for Emergency Decontamination

handles tend to absorb contaminants, all contaminated wooden items should receive special attention; discarding these items after use is usually the best decontamination practice. Respirators present special problems if regulators and nonrubber straps are contaminated by hazardous substances; harness assemblies and straps may have to be discarded after use. Regulators also require special maintenance by trained personnel after use and decontamination.

Heavy equipment, including backhoes, dozers, and front-end loaders, is difficult to decontaminate because of its bulk and weight. Accessible areas are traditionally scrubbed rigorously with brushes and detergent solutions and rinsed with high-pressure water to reach less accessible areas. Tires and other surfaces on large response equipment that is in direct contact with contaminants should be periodically checked by a swipe test.

9.9 GUIDELINES FOR PREPARING DECONTAMINATION SOLUTIONS

9.9.1 Solutions for Emergency Situations

Decontamination solutions are normally simple mixtures of water and chemical compounds designed to react with and neutralize specific hazardous substances and contaminants. Solution temperature and contact time must considered to ensure complete neutralization of hazardous agents has occurred. In some cases, firefighters and other response professionals may encounter a site containing unknown hazardous substances requiring that they receive complete decontamination after leaving the "hot zone." The following solutions should be used in situations with unknown hazardous substances because of their effectiveness for neutralizing a variety of contaminants:

- **Decon Solution A**—A solution containing 5% sodium carbonate (Na_2CO_3) and 5% trisodium phosphate (Na_3PO_4). Mix 4 lb of commercial grade Na_3PO_4 and 4 lb of Na_2CO_3 with 10 gal of water. These chemicals are available in most hardware stores.
- **Decon Solution B**—A solution containing 10% calcium hypochlorite ($Ca(ClO)_2$). Mix 8 lb of $Ca(ClO_2)$ with 10 gal of water. Calcium hypochlorite is commonly known as HTH and is available from swimming pool supply stores. Make sure HTH is purchased in plastic containers or is transferred from cardboard drums into clean plastic buckets marked "oxidizer."
- A general-purpose decon rinse solution for both decon solutions is a 5% solution of trisodium phosphate (Na_3PO_4) in 10 gal of water.

9.9.2 Decon Using Degradation Chemicals for Unknown Materials*

Five general purpose decon solutions are available for ten basic hazard classes:

- **Decon Solution A**—A solution containing 5% sodium carbonate (Na_2CO_3) and 5% trisodium phosphate (Na_3PO_4).
- **Decon Solution B**—A solution containing 10% calcium hypochlorite ($Ca(ClO)_2$).

* Reprinted from Michael S. Hildebrand. *Complete Decontamination of Personnel and Equipment Following Exposure to Hazardous Materials.* American Petroleum Institute and Prince George's County Fire Department, Hazardous Materials Response Team.

- **Decon Solution C**—A solution containing 5% trisodium phosphate (Na_3PO_4). This solution can also be used as a general-purpose rinse.
- **Decon Solution D**—A dilute solution of hydrochloric acid (HCl). Mix 1 pint of concentrated HCl into 10 gal of water. Stir with a wooden or plastic stirrer.
- **Decon Solution E**—A concentrated solution of detergent and water. Mix into a paste and scrub with a brush. Rinse with water.

The following can be used as a guideline for selecting degradation chemicals for a particular type or class of hazard identified:

1. Inorganic acids, metal processing wastes—A
2. Heavy metals: mercury, lead, cadmium, etc.—B
3. Pesticides, chlorinated phenols, dioxins, and PCBs—B
4. Cyanides, ammonia, and other nonacid inorganic wastes—B
5. Solvents and organic compounds such as trichloroethylene, chloroform, and toluene—C or A
6. PBBs and PCBs—C or A
7. Oily, greasy unspecified wastes not suspected to be contaminated with pesticides—C
8. Inorganic bases, alkali, and caustic wastes—D
9. Radioactive materials—E
10. Etiological materials—A and B

CAUTION: The decontamination solutions listed above are recommended for ten general groups of hazardous materials. Always contact expert assistance from manufacturers, poison control centers, medical specialists, and other professionals to determine the best solution to use.

9.10 REFERENCES

1. NIOSH, OSHA, U.S. Coast Guard, U.S. EPA. *Occupational Safety and Health Guidance Manual for Hazardous Waste Site Activities*. DHHS Publication No. 85-115, 1985.
2. U.S. EPA. *Hazardous Materials Incident Response Training Program*, 1983.

Safety in Confined Spaces

10.1 INTRODUCTION

Injuries and fatal accidents in confined work areas are very common for several reasons. Confined spaces and their accompanying hazards are not always easy to recognize, and people working in these environments tend to trust their senses. Many toxic substances, however, cannot be seen, smelled, or tasted. Contaminated areas of unknown hazard prevent the worker from staying alert to changing conditions in confined spaces, resulting in a false sense of area security and a higher risk of encountering an accident. More than 60% of the fatal accidents in confined spaces take place among unprepared rescuers; untrained and poorly equipped rescuers are usually injured along with those they attempt to help.

Confined spaces present a variety of hazards and require diligent and thorough safety procedures to ensure that the possibility of injury in minimized. Safe, proper entry procedures should be followed when entering a confined space. Furthermore, there should always be a rescue system in place, and a rescue should never be attempted without proper training, equipment, and backup. This section will discuss the various hazards associated with working in confined spaces and the appropriate safety measures required to avoid accidents. Confined spaces have the following characteristics:

- Few ways in or out
- Difficult to access
- Not intended for human occupancy on a regular basis
- Contain a hazardous atmosphere or other health or safety hazard

Examples of confined spaces include:

- Tanks
- Sewers
- Process vessels
- Utility vaults
- Pits
- Vats
- Boilers
- Manholes
- Pipelines
- Tank cars
- Trenches
- Barges
- Holds of ships

Although many jobs do not take place in confined areas, special situations, such as an emergency response, may require entry into confined spaces. Supervisors and safety officers often make decisions regarding those who work in confined areas; in these cases, managers should make sure all personnel understand the dangers as well as the proper safety procedures needed to safely access such environments.

10.2 RECOMMENDED PROCEDURES FOR ENTRY

Confined space entry consists of two critical elements:

- The space should be made as safe as possible for occupancy.
- In the event something goes wrong, a good rescue plan should be in place.

Important features of each of these elements will be discussed in more detail below.

10.2.1 Entry Permits

Entry permits are forms that ensure proper safety precautions have been taken before workers enter confined spaces. The following information is included:

- A description of the job being performed
- A report on the portion of the job already completed
- A list of special precautions needed to make the entry safe

Entry permit forms are issued for a particular space, time period, and purpose. The person entering should make sure all necessary procedures such as purging, ventilating, or isolating have been completed. Atmospheric testing, in addition, must be accompanied by the proper forms signed by the individual performing the test. Workers must always double check to ensure all required safety equipment, appropriate personal protective equipment (PPE), and special tools are listed in the entry permit and to verify a rescue procedure is in place.

10.2.2 Lockout/Tagout

One of the most effective ways to minimize confined space accidents is the lockout/tagout procedure. This procedure is designed to ensure that all electrical, hydraulic, pneumatic, mechanical, or other systems in a confined space are securely placed in a "zero energy state"—meaning these systems have been inactivated, locking *and* tagging all switches and actuating mechanisms. If machinery inactivation cannot be achieved by turning off switches, then the proper wires, coupling systems, chains, and belts must be removed. Also, pneumatic or hydraulic systems must have their pressure released prior to lockout and tagout. Once lockout/tagout has been performed, start-up should be attempted (locally and remotely) with all devices to ensure the zero energy state is in place.

10.2.3 Isolation Procedures

Isolation procedures are designed to protect workers in confined spaces from outside hazards, such as gases or liquids being pumped accidentally into a confined space, that could result in fatal injuries. The following are standard isolation procedures:

- Disconnect lines entering a space
- Insert a blank or blind in a line to block it completely
- Lock and tag valves and install a blank on line
- Use a double block and bleed

When blanks or blinds are used, a list of specific blinds that have been installed should be developed along with a list of the specific locations of all blinds that have been inserted. Marking blanks or blinds with colored tape or other indicator allows easy location after a job is completed. Another recommended practice is to use "T" handles on blinds; these prevent blinds from being mistaken for spacers.

10.3 AIR TESTING IN CONFINED SPACES

Most fatal injuries in confined spaces occur due to atmosphere hazards. Because many air pollutants cannot be detected by human senses, instruments must be used for prompt identification and elimination of toxic air contaminants. A wide variety of detectors are available for monitoring the three principal types of atmospheric hazards:

- Oxygen deficiency or excess
- Flammable gases or vapors
- Toxic gases or vapors

Detection of toxic gases is usually accomplished with direct-reading instruments—either gas detector tubes or electronic gas-sensing instruments. Many types of detector tubes have an accuracy of only 25–30%, which limits their usefulness for preliminary air screening procedures. Electronic devices are more accurate and therefore employed more frequently to detect high oxygen concentrations and flammable conditions. Some electronic devices can also measure the concentration of toxic gases such as carbon monoxide and hydrogen sulfide.

Instrumental air analysis should always be preceded by instrument calibration and testing area examination. Testing procedures must be optimized according to characteristics of the environment being analyzed:

- Volume of air
- Suspect pollutants present

Furthermore, atmosphere testing should proceed with the following considerations in mind:

- Testing space should not be disturbed.
- Test should be done from top to bottom, around ductwork and uneven surfaces.
- Gases could be trapped behind scale or sludge.
- Instruments should be rated safe for the atmosphere being tested.
- Time should be allowed for a sample to reach instrument sensors.
- Workers should avoid leaning over the space being sampled.
- When an instrument is being lowered into a space, readout displays should be clearly visible and alarm systems audible.
- Testing should continue until all detected hazards have been eliminated.
- Testing should occur frequently to make sure new hazards are not being created and old hazards do not return.

Our atmosphere is normally composed of about 21% oxygen. Concentrations below 19.5% are considered oxygen deficient, and levels above 25% classify as enriched. Oxygen loss in confined spaces can result from respiration, combustion, and other chemical processes or from displacement by other gases and vapors. Atmosphere enriched in oxygen is a concern because of the increased risk of fire.

Flammable gases or vapors must be within their flammable range to ignite. If gas or vapor concentrations are below the lower flammable limit, the air/vapor mixture is too "lean" to ignite (i.e., not enough fuel). If the concentration is above the upper flammable limit, on the other hand, the air/vapor mixture is too rich to burn (i.e., too much fuel). Air/vapor concentrations between the lower and upper flammable limit ranges may sustain combustion, provided oxygen concentrations are in the required range as well. Typical combustible gas detectors furnish readouts as a percentage of the lower flammable limit. These detectors are not designed, in most cases, to provide readings in the flammable range or above the upper flammable limit. Flammable gases, in addition, do not give accurate instrument readings if oxygen concentrations are below about 10–12%. Properly trained personnel, therefore, must recognize gas or vapor concentrations that fall off instrument scales and interpret such readings appropriately.

Concern about toxic gases usually arises at the parts per million or even parts per billion level. Because a universal gas/vapor detector has not yet been created, instruments must be designed with gas detector tubes fitted for each individual toxic gas. Although instrument accuracy decreases with increased sensitivity and specificity, an alarm from a particular instrument should never be ignored.

10.4 CONTROLLING HAZARDS IN THE ATMOSPHERE

There are a number of ways in which atmospheric hazards can occur in a confined space. Natural processes such as rusting and decaying of organic matter can remove oxygen and produce toxic or flammable conditions. Industrial processes such as welding, painting, cleaning with solvents, and fumigating can also produce hazardous conditions in a confined space. Methods designed to control generation of atmospheric hazards include good housekeeping, purging, and ventilation.

Mechanical ventilation is often used to circulate fresh air into a confined space. Ventilation may be used to bring air quality in work environments within acceptable limits for entry and work, as well as to remove harmful substances. Positive ventilation may be established by:

- Employing fans or air inductors, or
- Blowing air in or exhausting air out

In either case, it is important to achieve good air exchange and to bring air in from a good quality source. Under no circumstances should compressed nitrogen or oxygen be used for ventilation. In a flammable or dust-laden atmosphere, explosion-proof equipment, rated as safe to use in that atmosphere, must be used. Proper bonding and grounding should also be done to prevent static sparks.

Ventilation may not reduce hazards sufficiently, requiring workers to use personal equipment such as respirators and protective clothing. PPE must match the hazard(s) present. Air-purifying respirators filter or absorb contaminants and can be used under any of the following conditions:

- When oxygen levels are less than 19.5%
- When toxicant concentrations are too high for removal by filters
- When toxicants have no smell or taste
- When the exact toxic material is unknown or at an immediately dangerous to life and health (IDLH) concentration

Note: Fit testing is required for respirator use.

A self-contained breathing apparatus (SCBA) or airline respirator with at least a 5-min emergency escape bottle must be used when air quality cannot be kept at acceptable levels. These units must be of the positive-pressure type, have an air supply of breathable quality, be checked prior to use, be strapped on prior to entry, and should *never* be removed while in a confined space.

If the atmosphere in a confined space is IDLH, other protection is needed in addition to respiratory protection. This includes proper protective clothing if skin contact could be harmful, a full body or chest harness, and a lifeline retrieval system.

10.5 OTHER HAZARDS IN CONFINED SPACES

Hazards in confined spaces are not limited to the presence of chemical contaminants, flammable gases, or lack of oxygen. Entrapment may occur if a worker sinks in or becomes covered by loose material such as plastic pellets, sawdust, or other fine bulk material. Breathing passages can become clogged quickly in these situations; harnesses and lifelines are critical components of a successful rescue plan in these situations.

Machinery can create a number of hazards such as high noise levels, electrical shock, and entanglement with blades, rotors, or other moving parts. Unnecessary equipment should be disconnected or disengaged to prevent entanglement, and items that could shift should be blocked. To ensure all electrical equipment is in good repair, grounded or double-insulated tools should be employed. Ground fault circuit interrupters or low-voltage isolation transformers may also help to reduce the possibility of electrocution.

The potential for heat stress should always be taken into account when entering a confined space. Individuals should pace themselves, consume plenty of fluids, and be alert to the warning symptoms of heat stress. In addition, a monitoring program for individuals working under high heat load conditions should be in place.

Falls occur in confined spaces due to restricted space, dark and/or slippery conditions, and other reasons. Ladders should be set at the correct angle, be tied off at the top, have safety feet, and not be made of metal if work is being done around electricity. In situations where the potential for a fall is clear, fall-arresting devices should be used.

10.6 FIRST ENTRY

Initial entry into confined spaces should be done by qualified personnel intending to finish atmosphere testing and hazard evaluation. Protective equipment may include SCBAs, protective clothing, hard hat, radio, and lifeline. Typical hazards in addition to atmospheric hazards include unsafe footing; temperature extremes; low overheads or difficult accesses; biological hazards such as snakes, vermin, and insects; and ionizing or nonionizing radiation sources.

10.7 STANDBY PERSONNEL AND COMMUNICATIONS

The standby person is the critical link between those entering a confined space and the outside world. Also known as the "hole watcher," the standby person's foremost duty is to monitor workers in an enclosed area and to advise them of any hazardous conditions that may arise. This observer must:

- Know who is in the space
- Keep unauthorized persons out
- Maintain voice or visual contact with entrants
- Recognize early danger symptoms in the space
- Watch for hazards developing outside as well as inside the space
- Maintain clear access to and from the space

Standby personnel and workers in confined spaces should have a system of signals and visual communication. Depending on the circumstances, radio communication may also be required. In case of an emergency, standby personnel must:

- Be able to contact rescue teams
- Stay outside the confined space until their backup personnel arrives
- Perform rescue from outside the confined space where possible
- Assist rescuers and victims

Standard Operating Safety Procedures

11.1 INTRODUCTION

The health and safety of response personnel is the primary consideration when responding to accidental releases of hazardous substances or working at hazardous waste sites. While many technical operations must be performed in a timely and efficient manner to mitigate an emergency incident, the health and safety of the worker has to be given top priority. Safety-oriented operating procedures, in combination with appropriate equipment, trained personnel, and sound supervision and management, greatly reduce the risks response workers face.

For procedures to be efficient, they must be:

- Written whenever possible—Most can be written in advance of an incident or operation and then tailored to the specific situation.
- Based on the best possible information—Technical reports, papers, and other sources provide guidance of standard safety practices. Past experiences and personnel knowledge help make specific adjustments.
- Tested through training—Response operations and management systems must be practiced using realistic scenarios and hands-on exercises.
- Reviewed, critiqued, and revised—Through the use of practical exercises and drills and the careful review and analysis of actual incidents, procedures can be kept current and applicable.

11.2 RESPONSE ACTIVITIES

Upon arrival at the incident scene, responders should remain calm and in control. Check in with other responders to ensure your availability is recognized. If an incident commander or team leader has not already been assigned, designate who will be in command of the response operation. The commander, in turn, establishes a command post from which the operation will be directed and determines the response objectives. The scene should be controlled as soon as possible and the nature of the problem, along with any modifying factors, should be identified. Personnel should be advised of the chain of command and their specific assignments.

Depending on the nature of the incident and the operations that will be conducted, various personnel precautions may be enforced. Restrictions on eating, drinking, smoking, and so forth may specify that these activities take place only in designated areas and only after certain precautions are taken, such as thorough washing of the face and hands. In some situations, some of these activities may be banned for the duration of the response. Responders should remove all jewelry before donning personal protective equipment (PPE) to avoid punctures or snagging. Loose-fitting clothing and dangling straps, strings, etc. should be avoided if responders must work around equipment with moving parts. Hard hats, hearing protection, and steel-toe boots must be worn where appropriate.

If a responder has recently taken medication, the commander should be advised. If the medication could interfere with the worker's ability to work safely or could otherwise produce adverse health effects such as increased susceptibility to dehydration, the worker should be reassigned or told to stand down.

Above all, responders should be trained to avoid coming in contact with hazardous substances, whenever possible. Responders should walk around or step over puddles of hazardous substances if at all possible. When only one responder is required to perform a specific task that will result in contact with the chemical, the second responder should avoid contact while backing up and assisting the first. Pallets or other raised platforms to kneel on may prevent considerable contamination to the knees and legs. The use of absorbent pads or rags may prevent wiping contamination on protective clothing when gloves become slippery. The spill area, especially on small spills, may lend itself to being covered, or marked off, so that it is easier to avoid contact.

Ensure that personnel are adequately trained for the task they are expected to perform. It may be advisable to use personnel with different levels of expertise to make certain the operation is carried out safely. For example, a person familiar with tank car valve repair may not be familiar with the monitoring equipment that should be used; a safety person or instrument technician might need to accompany the tank car repair person. It is always advisable to practice the operation prior to actually entering the hazard area. Review who is going to do what, what equipment is needed, hand signals to be used, how the work will be done, and other aspects of the operation.

11.3 RESPONSE EQUIPMENT

Various methods can be employed to efficiently provide response equipment to the scene. Self-contained response carts, trailers, or vehicles help ensure that the equipment is accessible and available at the scene. Storage lockers, rooms, or buildings provide security and accessibility, but movement of equipment and materials to the incident scene may be more difficult. Whatever option is chosen, responders must be sure that response equipment is in good condition and of the right type and that a sufficient number will be provided in a timely manner.

Response equipment must be properly maintained and secured. A department or designated personnel can perform this function. In cases where response equipment must be shared, a checkout system should be used. Under no circumstances should response capabilities be jeopardized because of the need to share equipment. Adequate equipment and materials, in good working condition, should always be held in reserve for a response.

Before entering the hazard area, it is a good practice for responders to check their tools and equipment. Tool boxes should be opened to make certain the appropriate tools are there. A box should be packed so that one person can carry it. Only the specific tools and equipment necessary to do the job should be exposed to contamination. Do not contaminate the whole set of tools and equipment. Once contaminated, items should be kept separate until decontamination or disposal.

Proper management of the available breathing air supply is an area that frequently requires attention. Often air supplies are limited and replenishment or replacement is time consuming. Efficient management of air supplies makes for a safer, timelier response. Make certain that everything and everyone is fully prepared before anyone begins to breath down their air supply. This is more complex than it sounds. Responders, as well as their backups and the decontamination team, must be fully dressed out. Radios, monitors, equipment, and tools must be checked out and ready. The decontamination area must be set up. Responders should receive any briefings from the commander, safety advisor, or technical advisors before going on air. They must also determine their specific plan of action. Finally, emergency backup notification systems should be in place. Once all of these preparations have been made, responders are ready to begin using their air supply.

All entry personnel go on air at the same time. Backups do not go on air but are otherwise fully prepared for emergency entry. As a general rule, no entry should take place if air supplies are at less than three-fourths capacity. More conservative safety procedures may be desirable. Once on air, entry personnel should take full advantage of their air supply. Responders should move directly to the work area and perform the primary task assigned. They should be prepared to move on to other tasks as their air supply allows. Hot conditions, heavy physical labor, and other factors may call for limiting the work period to a specific, predetermined time period. Under less stressful circumstances, work time may be determined through frequent checks of the responders' air supply. Responders must cease operations and return to decontamination immediately upon notification that their allowable work time has expired.

PPE should provide maximum protection for both known and potential hazards. Standard chemical protective clothing should not be used for high heat locations, fire proximity work, or situations where conditions could result in a vapor flash. It is not designed for such use. Special flash-protective oversuits are available.

Whenever possible, it is advisable to use disposable or limited-use chemical protective clothing. This precludes the need for extensive decontamination in many situations and lessens concerns about permeation of the protective material and contamination of subsequent wearers. It is recommended that at least an inner and outer set of gloves be worn for any response. The use of outer boot covers is also advisable. This allows much of the gross contamination in the foot area to be removed with the boot cover rather than having to be washed and scrubbed.

An adequate number of personnel should be assigned to assist in the donning of chemical protective clothing. In situations where manpower is limited, responders and their backups may have to be able to assist each other. Be sure to check responders before allowing them to enter the hazard area. Check to ensure that air supplies are within acceptable limits, all openings are closed or taped over, and that the responders are physically and mentally capable of performing their assigned tasks. Look for signs of physical and/or mental fatigue or stress, and do not hesitate to have personnel stand down if symptoms are observed.

11.4 APPROACH AND ASSESSMENT

Exercise caution when the chemical hazards are unknown or unidentified. Do not assume that the chemical(s) has been identified until the identity is confirmed by the response team. Approach from the upwind side. Do not overlook the fact that spills may be moving downslope or downstream. The route a spill may take via below-ground drainage systems should also be a consideration during the initial approach and set up. Binoculars, spotting scopes, and zoom lenses are useful for reading labels and remotely identifying the leak source. Witnesses and personnel who work in the spill area may be able to provide important information. It may be helpful to diagram a map of the area, noting points of reference to avoid possible confusion in planning and communicating.

Monitoring should be done during the initial approach by entry teams, especially when there is a possibility that flammable vapors or gases are present. Personnel using the monitors must understand the operation of the instrument and the action levels that have been established. Personnel entering initially should be protected to a level that will provide for their safety at the maximum toxic concentration that could be encountered. As toxic action levels are detected, they should be marked or the information communicated back to command. If flammable action levels are reached, entry teams should pull back, unless properly protected for this hazard. PPE typically used for chemical response or firefighting is not designed to protect workers from the blast of an explosion; therefore, extreme caution must be taken when possibly explosive conditions exist.

Monitoring may take many forms. It is extremely important to characterize the work area through initial toxic, oxygen level, and flammable vapor testing. This will assist in fine-tuning work zones and establishing levels of protection for subsequent entry teams. In situations where conditions could change rapidly without warning, continuous monitoring is recommended. In other cases, monitoring may be done periodically, such as when it is determined that changes in ambient conditions or specific operations could increase chemical concentrations. Monitoring can also help determine when an area is "clean" and how effective decontamination has been.

11.5 SITE CONTROL

Effective site control is essential for the safety of response personnel, as well as other persons who may be in the area. It begins with effective notification procedures for the area involved and other areas that may soon be affected. Notification is followed by timely and orderly evacuation of nonessential personnel to a safe area and response personnel to a designated muster point.

A "hot zone" perimeter should be established by the incident commander. This area is then isolated and access restricted to controlled entry and exit points. Initially, the hot zone boundary is determined based on the nature of the chemical involved, the type of container affected, and the location of the incident. As reconnaissance and monitoring are done, this boundary is subject to change. Entry into the hot zone requires approval from command, along with the use of specific PPE, which must be specified. Time of entry and number of persons entering must be noted. Emergency evacuation signals should be understood and other methods of remote communication checked out prior to entry.

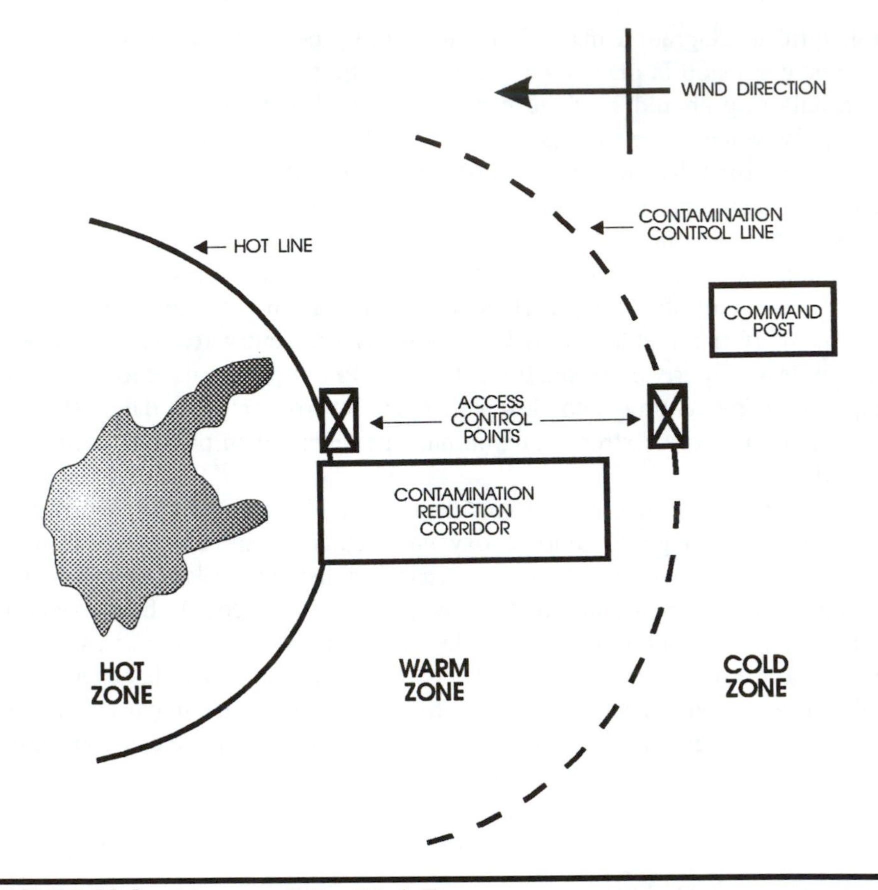

FIGURE 11.1 Diagram of Site Work Zones

In addition to the hot zone, a warm zone and a support zone should be established. The warm zone serves as a safety buffer area just outside the hot zone (see Figure 11.1). Within the warm zone is the contamination reduction corridor. All personnel exiting the hot zone who have come into contact with the hazardous substance should pass through the contamination reduction corridor. This allows control of the physical transfer of contaminating substances on people, clothing, and equipment. The cold zone is considered a noncontaminated or clean area and is the outermost area of the operation. Potentially contaminated personnel, clothing, equipment, and materials are not permitted in this area but are left in the warm zone. The cold zone contains the command post, along with response equipment, materials, and personnel who are involved in the operation.

11.6 ENTRY TEAM SAFETY

Personnel entering the hot zone should use the buddy system. This means that a minimum of two responders enter at one time. An equal number of backup personnel, in the same level of protection, should be on standby for emergency assistance or rescue. A briefing should be given by the safety officer and the operations chief prior to entry. This should cover essential safety considerations and dangers, along with the operational objectives. A radio check should be made and the air supply and time of entry noted. Visual contact should be maintained if at all possible. Audible and visual signals may be used for emergency notification in the event of radio failure.

A decontamination area must be established and decontamination personnel prepared prior to entry whenever contamination could occur. Initially, this area is considered clean, but will "warm up" as decontamination activities commence. Clearly define the decontamination corridor with marking tape or other markers so that personnel exiting the hot zone will not have difficulty staying within its bounds. This will also prevent inadvertent entry by personnel from the support zone. Do not overlook the need for addressing emergency decontamination procedures.

Medical surveillance begins with baseline physicals conducted prior to response work. Additional examinations are called for whenever harmful levels of exposure occur or are suspected. At the scene, it is vital that personnel be monitored for signs of heat and cold stress and that appropriate steps be taken to protect workers. Medical personnel, as well as safety personnel, supervisors, and co-workers, should be alert to signs of psychological and physical stress in workers.

Emergency medical personnel should be placed on alert and advised of potential problems whenever a chemical emergency response occurs. Ambulance crews should be on standby. It is unwise to assume medical personnel are familiar with specific types of chemical exposures and treatments. Material safety data sheets and specific toxicological information may need to be provided. Injured persons who are contaminated should be decontaminated prior to transport, if the decontamination procedures will not compound the injury.

The safety officer is responsible for the implementation, revision, and enforcement of the safety plan and the standard safe operating procedures. This person recommends the work zone boundaries and ensures that medical surveillance and emergency care and transport are available. The safety officer also makes final decisions on entry/nonentry, withdrawal, evacuation, levels of personnel, protective equipment, monitoring protocol, and decontamination procedures.

12

Hazardous Chemical Spill Containment

12.1 INTRODUCTION

Spill containment may be the most important step in controlling hazardous chemical releases. However, response personnel generally have little formal knowledge regarding available equipment and strategies with which to competently manage containment efforts. Effective spill containment should protect watercourses, drainage systems, and sensitive environmental areas through timely and cost-effective application of containment techniques. In many cases, initial containment options are limited by available materials, equipment, and personnel. Two major points should be considered when developing initial spill containment strategies:

- Protection of emergency response personnel
- Possible detrimental environmental effects from containment application

Containment activities may require response personnel to come in close proximity with spilled material. Proper protection and safety precautions such as protective clothing, breathing apparatus, decontamination procedures, and backup personnel must therefore be included as part of the final containment strategy.

Once the potential environmental effects of a spill have been evaluated and appropriate safety precautions have been taken, containment efforts can be initiated. Final choice of the most appropriate containment technology is dependent on a number of factors:

- Spill volume
- Spill location
- Class or type of material spilled

Hazardous chemical releases may generally be classified as land or water spills. The following sections describe some available techniques for spill containment in each type of location.

12.2 LAND SPILLS

Hazardous chemical releases that occur on land do not travel as far or spread as rapidly as spills into water resources because soil, vegetation, and other materials tend to restrict the lateral movement of chemical mixtures. Liquid can exhibit rapid vertical movement in porous soils, however, resulting in the contamination of large volumes of soil and possibly groundwater resources. Hazardous chemical spills in urban areas may also flow into storm sewer systems or utility excavations, causing contaminants to spread and making removal difficult.

Spill containment on land is accomplished primarily by:

- Confining the spill in some form of excavation
- Erecting dikes
- Erecting soil or solid barriers

Dikes may be constructed from on-site (dirt) or commercially available materials (cement). Liquid wastes that have been spilled can be collected in exiting depressions, ditches, or pits. It may be necessary in some cases to dig designated channels to allow spill movement into natural depressions or specially constructed pits. When possible, holding pits should be lined with an impervious material; if collected waste materials are noxious or produce flammable vapors, holding pits should be covered. Many types of fire-fighting foams or special hazardous material foams can be used to provide satisfactory vapor barriers in these pits.

Excavation is not a complicated containment method, but several factors need to be considered:

- Suitability of soil for excavation
- Soil porosity
- Amounts of material to be moved
- Spill volume and available area for pit excavation
- Available equipment and personnel

Land spills can also be contained by building physical barriers that divert hazardous chemicals away from sensitive environmental or ecological areas or completely confine released substances. However, because the materials used to build land barriers may absorb spilled chemicals, care must be taken to minimize the volume of final ditch material that will require long-term storage or disposal.

A growing number of commercial devices and materials that may be used to confine or divert a spill are available. Several manufacturers produce small portable chemical spray systems, which supply a quick-setting polyurethane foam to a widely dispersed chemical spill; this foam solidifies in less than a minute and adheres readily to hard surfaces such as asphalt or concrete. Corn starch–polyacrylonitrile graft copolymer is another spill-control material used at emergency sites. This mixture, which forms a sticky paste when mixed with water, comes premixed as well as preapplied to a mesh backing; this "matted" form is highly suitable for sealing grates and drains. Other specially engineered matting materials include pliable elastomers (rubber-like materials) used as a protective cover for drains, grates, or manholes. Soil or other on-site materials can be used around and over these synthetic polymer sheets to ensure a good seal is formed in order to minimize hazardous chemical movement.

12.3 HAZARDOUS SPILLS ON LARGE WATER BODIES

12.3.1 Introduction

Spills entering a waterway can be described as floating, sinking, or water soluble in nature; containment techniques, consequently, must be optimized with these properties in mind. Hazardous substances entering a waterway are subject to the same movement and dispersal forces that affect the body of water. Because water movement can cause spills to travel rapidly over wide areas, proper containment methods are critical.

Containment of substances that sink, furthermore, may be compounded by the inability to visually observe spilled materials. Current speed can be used in this case to estimate how far the material has moved. In some types of waterways, such as tidal streams and irrigation canals, spills can be contained by controlling water flow (irrigation canals use gates and pumps to move water). Streams, creeks, and other slow-moving waterways may allow for the construction of trenches or depressions ahead of a spill, as shown in Figure 12.1. Chemical releases in fast-flowing, deep bodies of water are not easily contained by this method.

Water contaminated with soluble chemicals or suspended materials may require containment or isolation as a separate body of water. A containment method often employed in this situation is to construct a dam upstream and

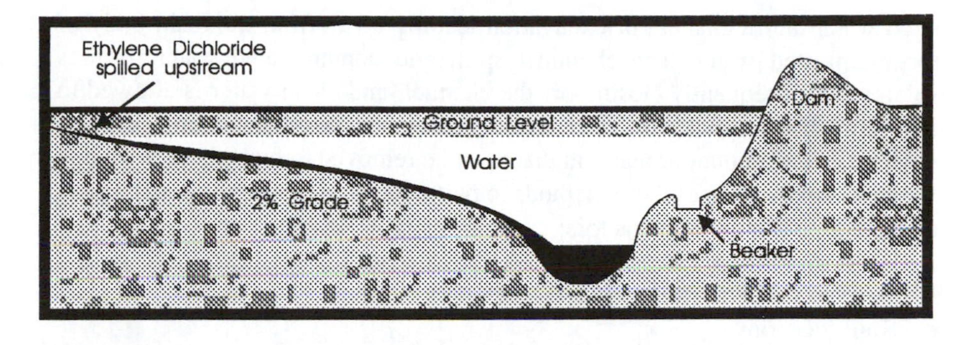

FIGURE 12.1 Containment Trench for Heavier-Than-Water Chemical Spill

downstream of the spill area. Pumps and portable plastic piping can subsequently be used to divert water stream flow around the contaminated area, as shown in Figure 12.2. If this containment method is used, large pumps and considerable amounts of pipe are often needed to properly bypass contamination from the immediate spill area.

Soluble chemical spills may also be contained by constructing diversion channels. This containment method employs dams as well as a channel constructed around the area containing contaminated water to shunt clean water away from the site. A useful modification to this approach is to divert contami-

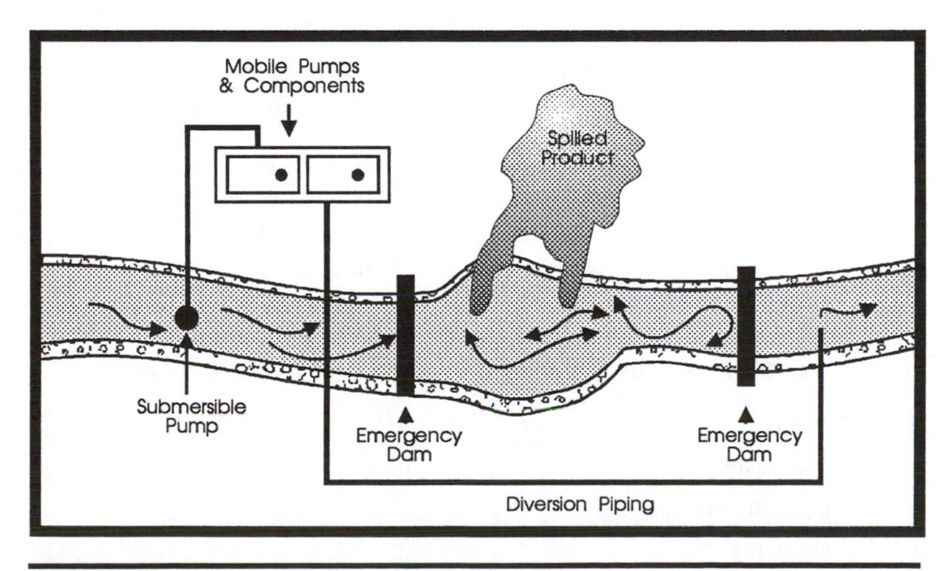

FIGURE 12.2 Stream Diversion System

nated water into a channel or excavation leading away from the main body of uncontaminated water. The chemical spill, and contaminated water, is thus isolated by subsequently closing off the channel, and clean water is allowed to flow through its normal path. The use of dry creek beds or other existing channels may reduce the amount of material that must be removed to form these diversion channels. Application of these methods to particular situations and sites is highly dependent on a number of factors:

- Watercourse characteristics
- Spill location
- Contaminated area identification
- Availability of equipment
- Promptness of containment initiation

Methods employed in efforts to contain hazardous chemicals that float are more specialized. Technologies have been developed over a number of years to properly manage and contain oil spills; these same methods and equipment are directly applicable to chemical spills that float.

12.3.2 Mechanical Booms

Mechanical booms are the most efficient tool presently available to contain floating spills. Booms consist of:

- An upper freeboard section, which incorporates polymeric foam or air chambers to provide flotation and surface containment
- A lower section called the skirt, which prevents subsurface loss

The skirt is usually composed of a fabric fence supported by chain, lead weights, cable, or the stiffness of the fabric material. Tension members are often incorporated to assume the load from the water pressure acting on the boom and to help maintain the boom in an upright position (see Figure 12.3).

Booms are commercially available in a wide variety of sizes and are composed of a wide variety of materials. Overall boom height can range from 6 in. to more than 6 ft. Common fabric types used in boom construction include:

- PVC/nylon
- PVC/polyester
- Neoprene/nylon
- Polyurethane
- Polyamide/PVC

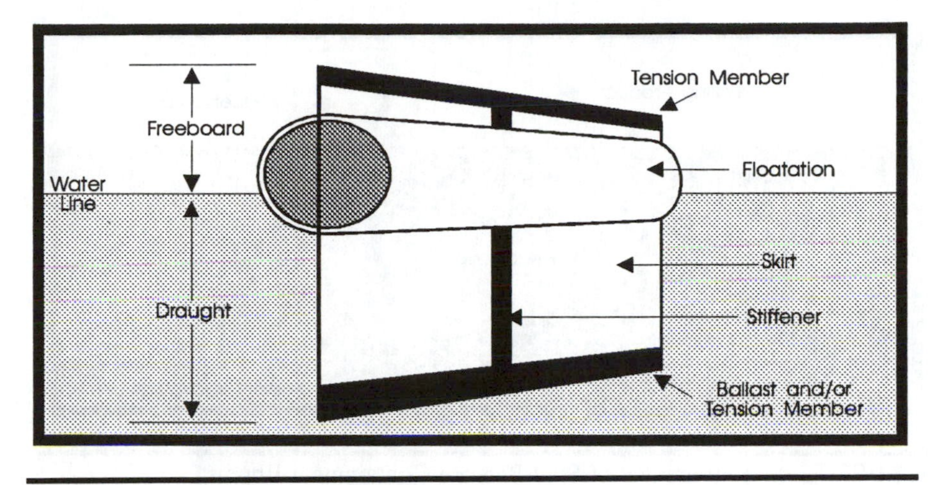

FIGURE 12.3 Boom Design Features

Over time, these fabrics may be chemically degraded to the point of structural failure. Fabric contact with corrosive and hazardous chemicals is usually restricted to the waterline area due to normal movement (waves, splashing). When containing spills in rivers and near-shore waters, boom height does not traditionally need to exceed 20 in. Typically a boom is selected with a freeboard of at least 30% of the wave height. In calm waters, almost any boom proves effective, although faster currents and choppy water may require booms with higher freeboards and round flotation members.

Boom containment of spilled chemicals should prove most effective in currents ranging from 0 to 1.5 knots. Above 0.7 knots, chemical or waste containment becomes more difficult because floating material may begin to entrain beneath the boom, as shown in Figure 12.4. Chemical entrainment can be reduced and controlled by angle deflection and by using double booms. Figure 12.5 illustrates the angles of deflection needed to prevent chemical entrainment as water current speed increases; by decreasing the boom deployment angle, the load on the boom per unit area is reduced and entrainment becomes less pronounced. Angle deflection is normally employed in conjunction with double booms. The position of the second boom should be just beyond the point where entrained material is resurfacing. Some situations require a third or even fourth boom to contain spilled material and deflect it into shore for recovery.

Structural failure of containment booms may also occur due to splashover if precautions are not taken to ensure that height and reserve buoyancy are sufficient for the encountered wave action. Boom leakage, in addition, can result when contained material builds up to a depth greater than the boom skirt, causing leaks to develop underneath the boom. Both of these boom failure scenarios can

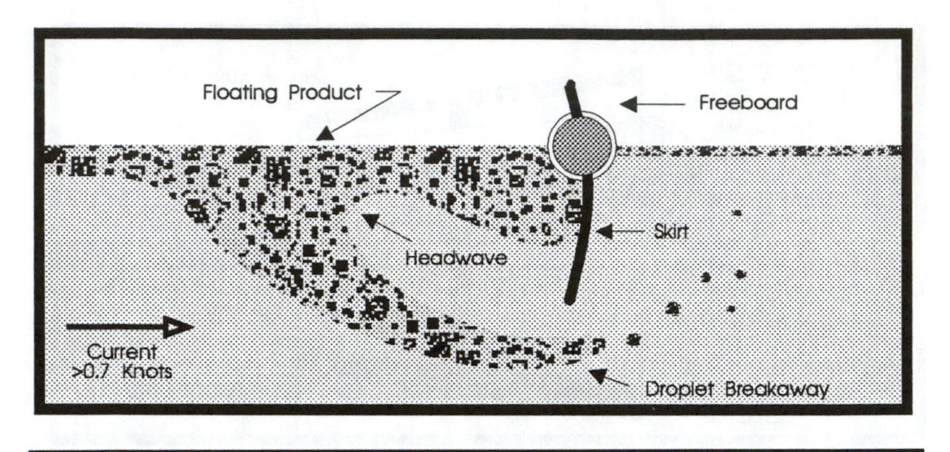

FIGURE 12.4 Entrainment of Spill Beneath Containment Booms

be prevented through proper planning during design of spill containment. The use of booms for spill containment has several important limitations:

- Once deployed, booms may impede navigation of surface transport vessels.
- Booms may not be available in a timely fashion when spills occur at remote locations.
- Booms tend to overload and fail due to floating debris that gets collected along with the spilled material.

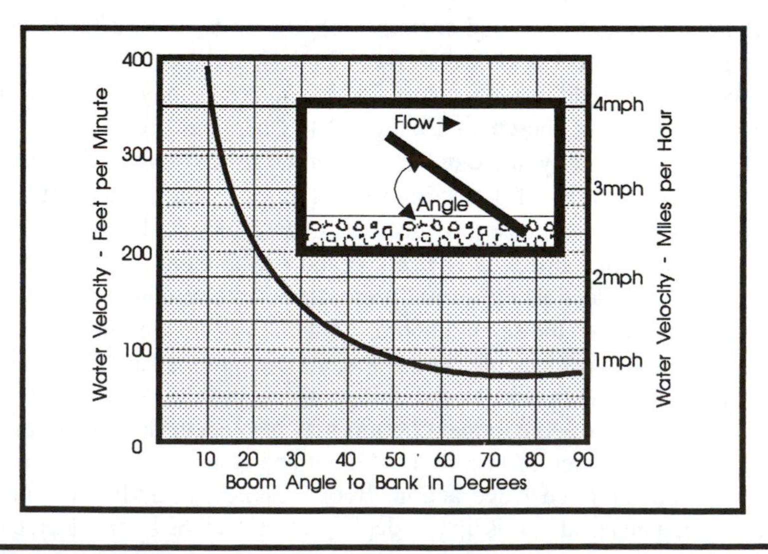

FIGURE 12.5 Deflection Angles for Containment Booms at Various Water Speeds

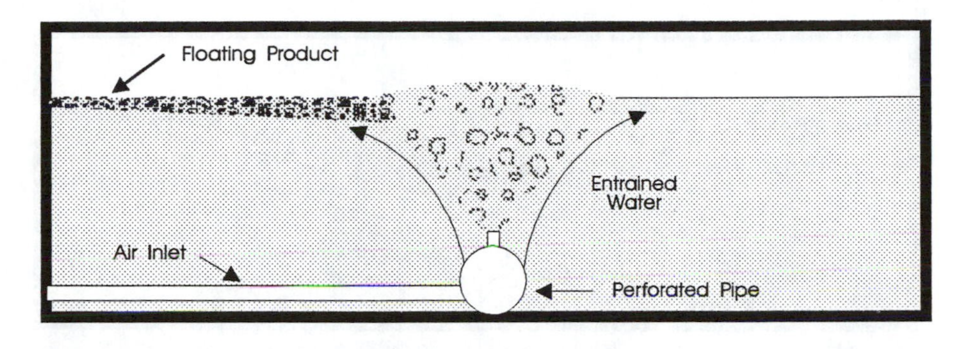

FIGURE 12.6 Pneumatic or Bubble Barrier Containment System

12.3.3 Pneumatic Barriers

Pneumatic or "bubble" barriers create a curtain of bubbles and turbulence capable of containing floating spills in low-current situations (less than 0.5 knots). Figure 12.6 illustrates a model pneumatic barrier for containing a chemical spill. Large-volume air compressors, hoses, and perforated manifolds are required for this method. Curtains of bubbles do not represent a hazard for navigation but cannot avoid the same problems booms encounter with large floating debris. This method is not as widely used because of equipment requirements and limited effectiveness in moving water streams.

12.3.4 High-Pressure Water Systems

High-pressure water systems rely on a battery of floating nozzles and a high-pressure pumping system to produce jets of water. The high-pressure jets of water are directed downward and at an angle to the water surface, creating turbulence and a headwave. Floating spills that encounter this turbulence are deflected and contained. These systems are currently in the research and development stage, but may find future applications in hazardous spill containment.

12.4 CONTAINMENT OF SPILLS ON SMALL WATERWAYS

A considerable number of hazardous material spills occur on small waterways including creeks, streams, and ditches containing runoff waters. Commercial booms are often too large or unavailable to effectively service these spills, but a number of other control measures tailored to small waterways can be effective.

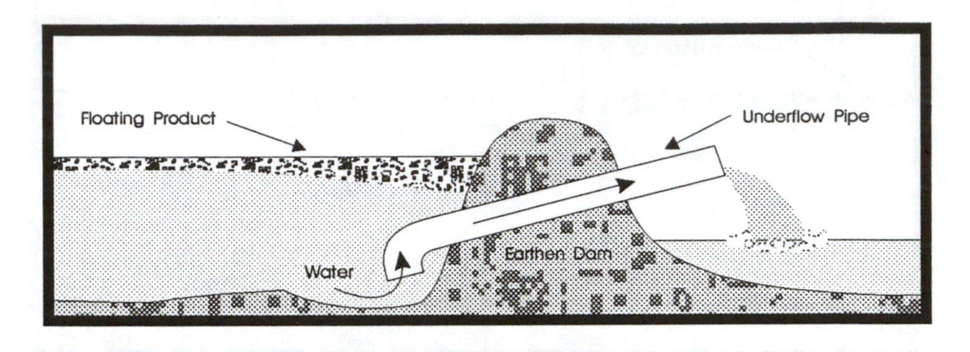

FIGURE 12.7 Diagram of an Underflow Dam

12.4.1 Underflow Dams

Underflow dams, as shown in Figure 12.7, are constructed of a series of pipes that control subsurface water flow while surface spill material is maintained behind a soil dam. The underflow pipes must be of sufficient diameter and number to prevent water build-up and eventual overflow. The upstream end of the pipes should be angled downward so that water will be drawn from far enough below the surface to prevent floating spill material from being pulled through the pipes. Several techniques are used to prevent vortex formation and subsequent spill loss. Laying a piece of plywood over the area will reduce this effect. In addition, the upstream pipe ends can be cut at about 45° angles so that portions of the pipe extend over the water intake area. Elbow-shaped pipe can also be used to lengthen the water intake to a lower level and thus eliminate vortex action.

Soil around an underflow dam should be compacted tightly around component pipes because water finds outlets through the dam along the pipes. Pipe ends on the downstream side of the dam should be far enough away to prevent partial or total washout of dam material; periodic monitoring of dam structures is required because silting can occur in and around pipes, causing the water level to rise.

12.4.2 Overflow Dams

For overflow dam systems, as shown in Figure 12.8, a pump or series of siphon hoses is used to pass flowing water over the top of the dam while containing the floating spill. Pumps or siphon hoses must be present and capable of moving sufficient volumes of water to prevent overflow and contain spills that float or sink.

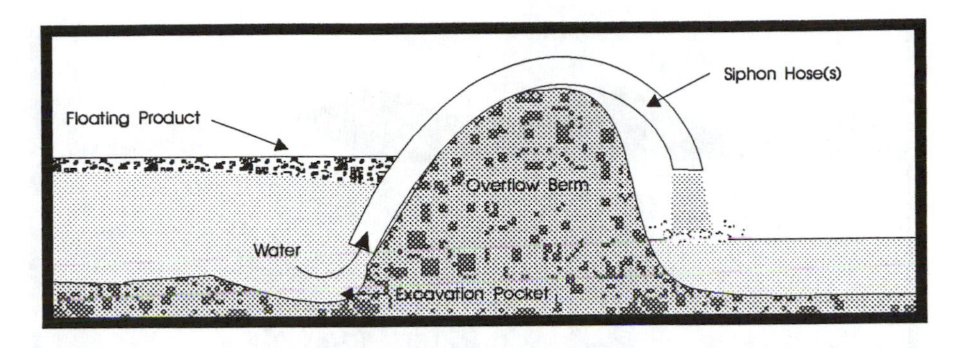

FIGURE 12.8 Overflow Dam Containment Mechanism

12.4.3 Containment Weirs and Sorbent Filter Fences

Weirs are impermeable "gates" placed so that water can pass underneath while preventing floating chemical contaminants or spills from moving and dispersing away from the spill site (Figure 12.9). A weir can be constructed using any of the following materials:

- A piece of plywood
- Solid lumber such as a 1 × 6-in. board
- Plastic sheeting over chicken wire

Handles can be attached so that the weir can be easily raised or lowered. These structures are subject to the same entrainment problems that affect booms,

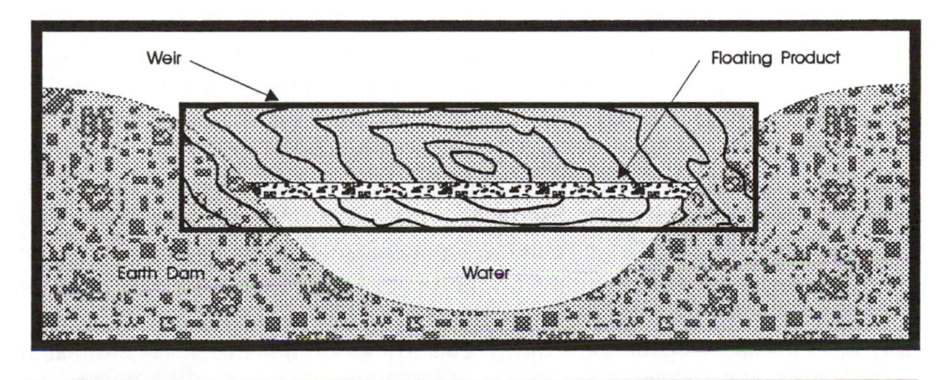

FIGURE 12.9 Containment Via a Weir

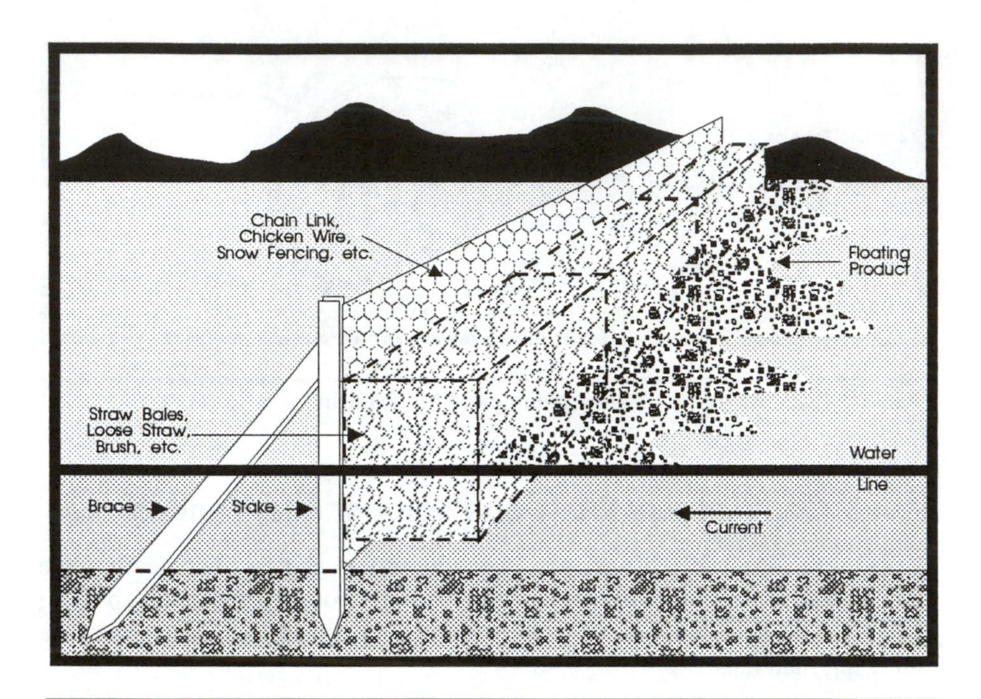

FIGURE 12.10 Straw or Sorbent Filter Fence

and the remedies suggested for booms work equally well in the maintenance of weirs. Angled deflection of water into a sump trenched along the waterway shoreline, for example, often allows weirs to more effectively contain spilled chemicals or other hazardous materials while allowing uncontaminated water to drain off site.

Sorbent filter fences (Figure 12.10) are another coordinate technology that has proven effective in containing spilled hazardous substances. Filter fences employ a sorbent material, usually in loose form, to absorb or filter out floating contaminants while allowing clean water to pass underneath. Fence material or netting, placed on the upstream side or between two pieces of netting, is used to secure the sorbent. The resulting "sandwich" effect is useful for creating a thick bed of sorbent material. The absorbent, however, must be replaced frequently after becoming saturated with contained contaminants. Several materials may be employed as sorbents, including straw and commercial loose sorbents. Straw is not recommended for long-term containment, however, because it will eventually absorb water and sink. Straw also has a low absorption efficiency compared with commercial sorbents.

Patching and Capping of Leaking Containers

13.1 INTRODUCTION

Patching and capping refer to the application of devices or materials to temporarily repair containers leaking hazardous chemicals. Pressurized containers may develop leaks around valves and fittings; such leaks are generally eliminated or "capped" with special leak repair kits. Patching, a similar procedure, refers to the use of various materials to seal or plug openings in low-pressure containers.

13.2 CAPPING

Capping kits are designed according to the valve system they will service. Container designs and valve arrangements are not consistent for all chemicals, and a capping kit for a particular type of container may not be adaptable to valve systems for other chemicals. Capping systems, furthermore, are available for only a few valve systems. Therefore, servicing a high-pressure container may require proper planning. Figures 13.1 and 13.2 show capping kits for chlorine containers.

Capping should proceed only after the source of a leak has been clearly identified and attempts at sealing the leak by other means, such as tightening valves or fittings, have failed. At least two people are required to apply a capping kit, and the container must be in a position to receive the capping device. The following are important considerations to keep in mind when attempting capping:

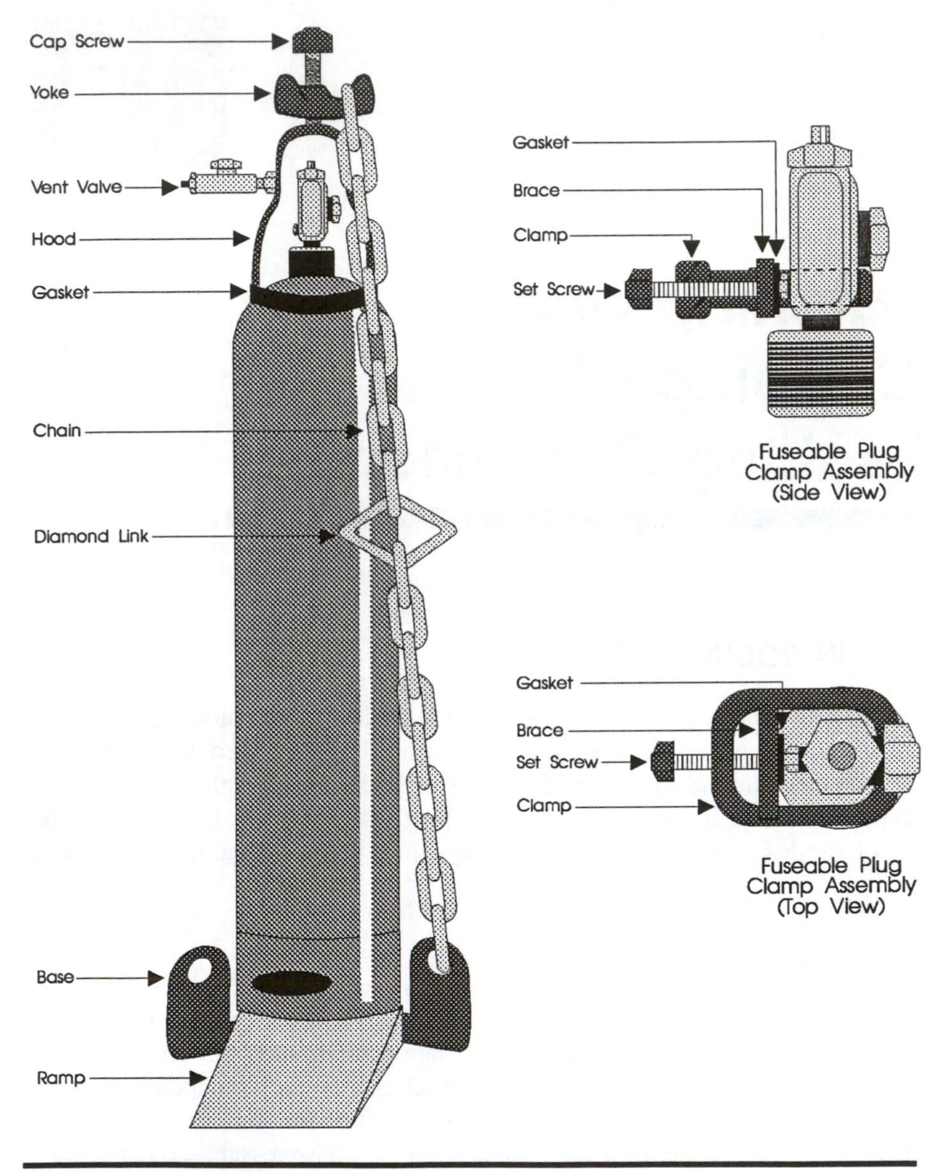

FIGURE 13.1 Emergency Kit for 150-lb Cylinders

- Caps must be vented while being installed to permit formation of a tight vapor seal around the gasket.
- Pressure caps have limited pressure capacity, and extreme pressure build-up may shorten cap lifetime.
- Cap and gasket material may weaken or corrode by contact with the product, leading to unexpected failure.

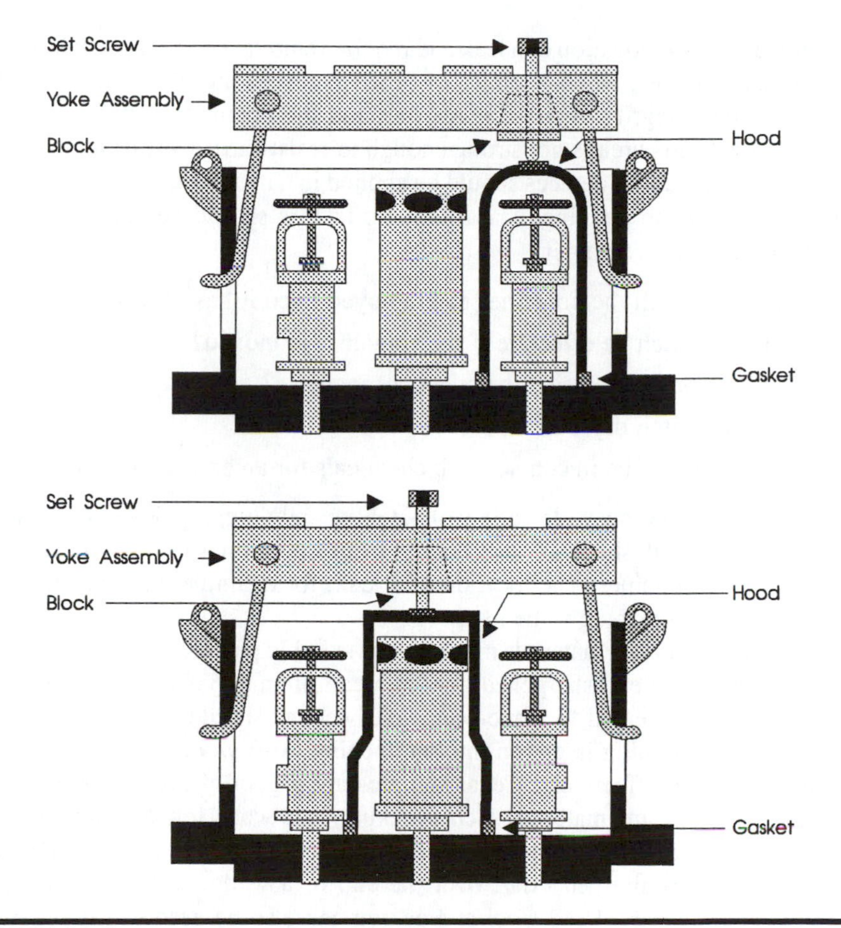

FIGURE 13.2 Emergency Kit for Railroad Tank Cars and Tank Trucks

- Corrosion, rust, and pitting around the pressure plate may prevent the cap gasket from sealing properly. Scraping and/or application of epoxy may help to provide a smooth surface and allow a good seal.
- After a kit has been applied to a tank car, the hatch cover must remain open. This may cause problems in moving the car because of the added height.
- To be effective, gaskets must be of the proper size and configuration and be in good condition.

13.3 PATCHING

Patching is a relatively simple procedure, although care should be taken so that patching devices are not rapidly deteriorated when applied. Insulation may disguise the exact location of a leak and require removal before a patch is applied. If wooden wedges, wooden plugs, or other materials are driven into a leaking opening, the integrity of the container around the opening should be carefully monitored. If this area is not strong enough to resist further pressure as a wedge is driven in, patching devices should be applied in a manner that does not enlarge or further tear the opening. The following factors should be considered when patching a faulty container:

- How long will the container be employed once it has been patched?
- Will the patch be effective if the container is moved?
- Will the pressure inside the container increase?
- Will the patch dry out?
- Will the patch be in contact with chemicals for an extended period of time?

Having access to a wide variety of patching and plugging devices is desirable because size and shape of leaks vary from situation to situation and from container to container. The location of a leak, for example, may determine the particular patch or plug to be used.

Many objects and materials may serve as suitable patches or plugs (see Figure 13.3). For example, a sharpened tree limb can often serve as an effective plug. Ideally, wood selected to make plugs and wedges should be soft to allow for spreading as the plug is driven into an opening. Soft woods, such as redwood, seal leaks more effectively. Cedar shingles are a good source of prefabricated wooden wedges, and materials such as oakum, lead wool, felt, and neoprene may be used to plug small punctures. A tight fit can be obtained by placing oakum, lead wool, or similar materials over the end of a wedge and then driving the wedge into the leak. Lead wool and oakum can also be used to fill small gaps after a plug or wedge has been used.

Various types of screws, T-bolts, and toggle bolts may also be used to fabricate plugs. Rubber balls and rubber stoppers provide good sealing capability when compressed on a toggle bolt or T-bolt. Pieces of neoprene in several thicknesses may be cut to fit various combinations of screws and washers. Hardware stores and plumbing supply shops have many items suitable for patching including various types of epoxies, plumbers' plugs, self-tapping screws, and the other items mentioned above.

Commercial patching devices and patching kits are available from a number of manufacturers. Vetter Systems® manufactures inflatable leak-sealing bandages and inflatable plugs for sealing pipes and other circular openings. Commercial patch kits combine various types of patching materials with the tools needed for applying the patches. Typically, neoprene plugs, wooden plugs, T-bolt/rubber stopper combinations, sheets of neoprene, and various types of putty may be found in these kits. Examples of patching techniques are provided in Figure 13.3.

FIGURE 13.3 Techniques and Equipment for Patching and Stabilizing Leaking Containers

Emergency Response Plugs

SOFT WOOD WEDGE

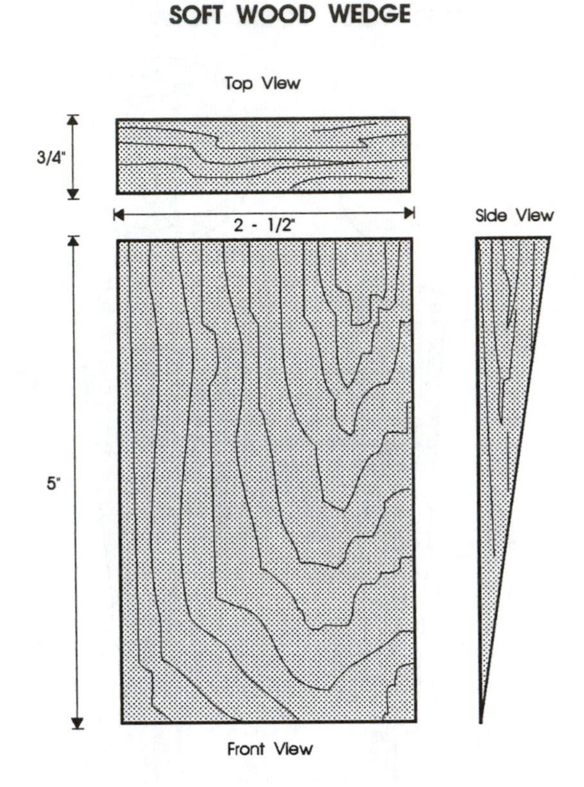

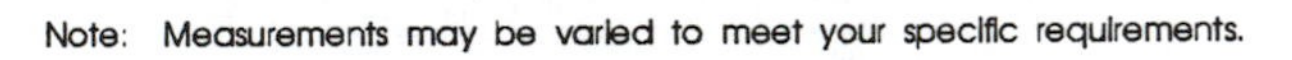

FIGURE 13.3 (continued) Techniques and Equipment for Patching and Stabilizing Leaking Containers

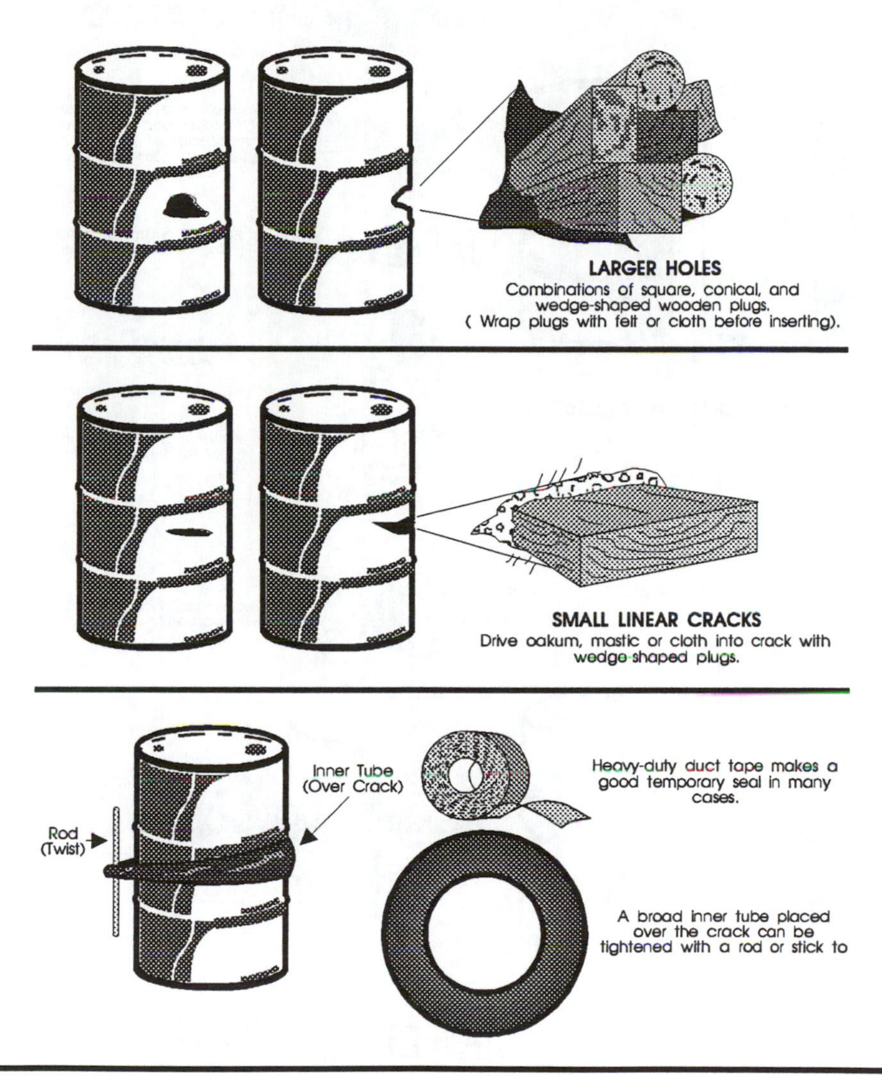

FIGURE 13.3 (continued) Techniques and Equipment for Patching and Stabilizing Leaking Containers

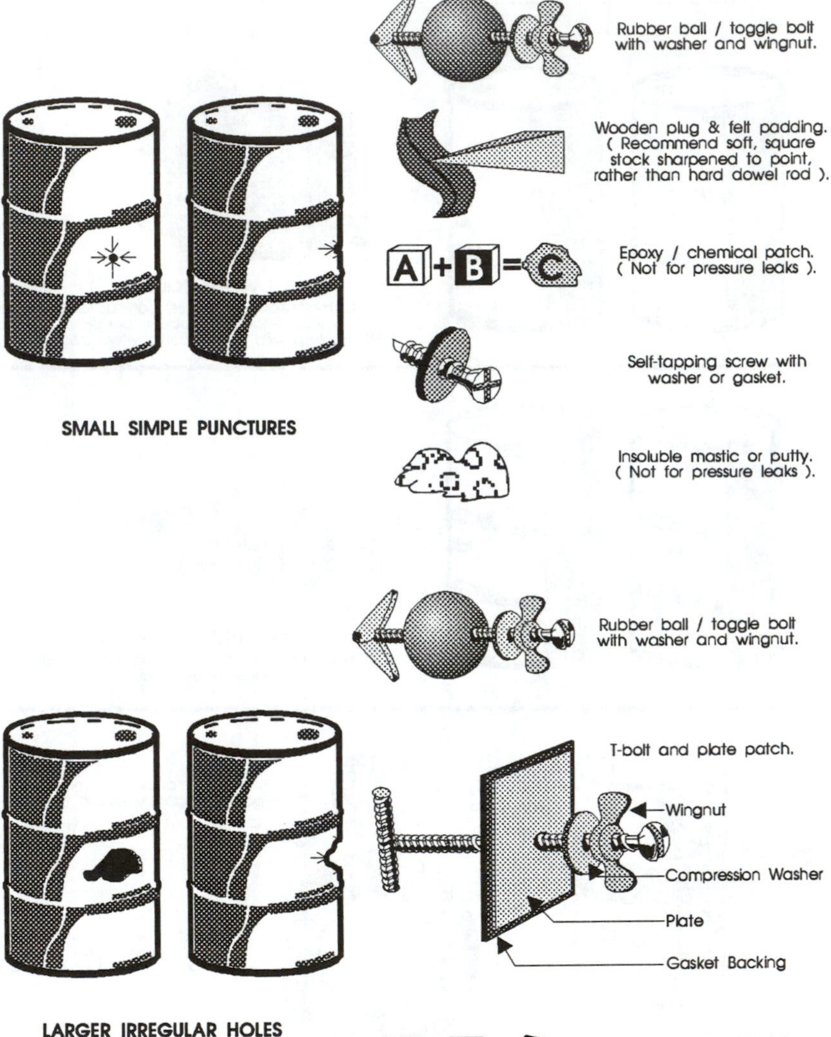

FIGURE 13.3 (continued) Techniques and Equipment for Patching and Stabilizing Leaking Containers

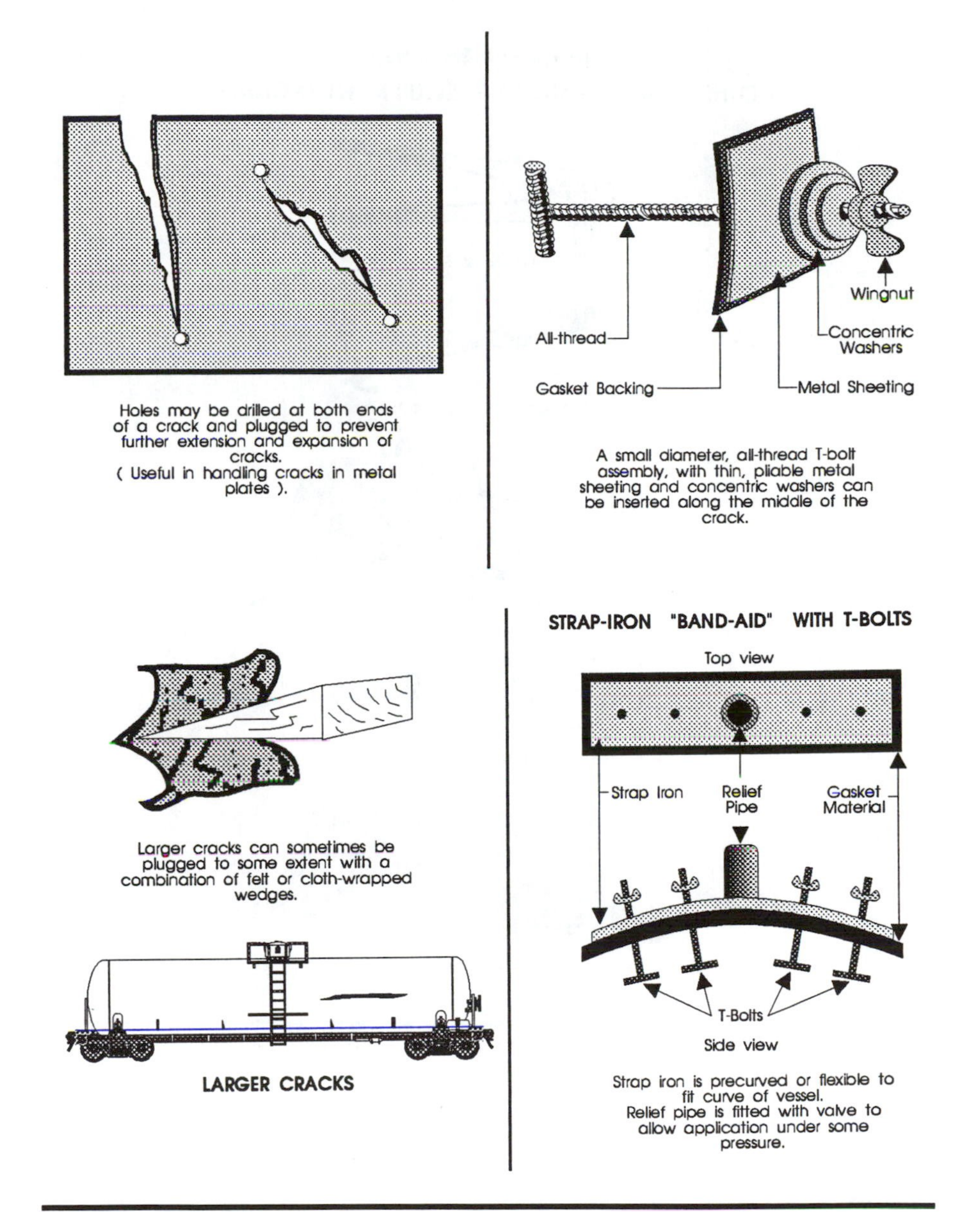

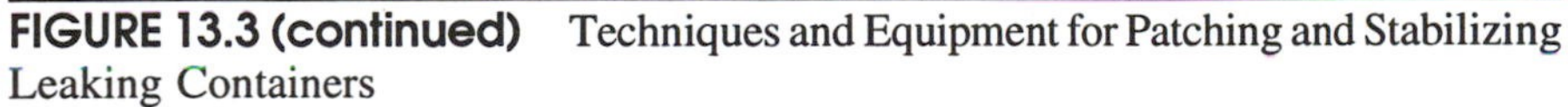

FIGURE 13.3 (continued) Techniques and Equipment for Patching and Stabilizing Leaking Containers

Emergency Response Plugs

OTHER USEFUL PATCHING EQUIPMENT INCLUDES:

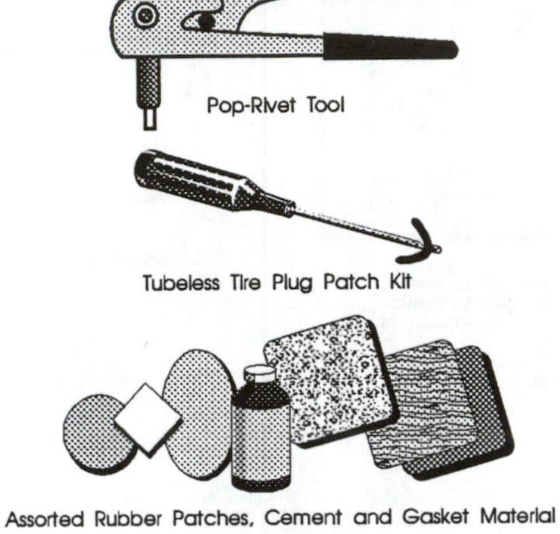

Assorted Rubber Patches, Cement and Gasket Material

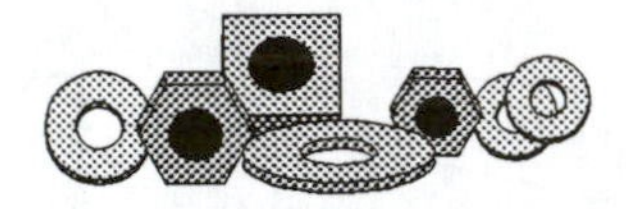

Assorted O-Rings, Washers and Nuts

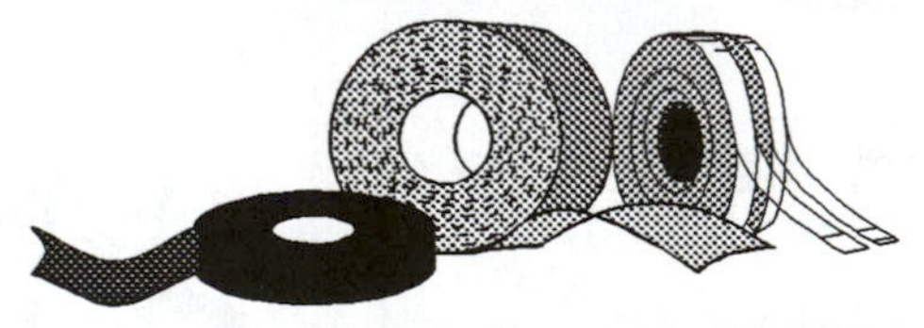

Various Tapes - Duct Tape, Teflon Tape, Electrician's Tape, etc.

FIGURE 13.3 (continued) Techniques and Equipment for Patching and Stabilizing Leaking Containers

14

Techniques for Spill Stabilization

14.1 INTRODUCTION

The accidental release of hazardous materials and substances into the environment often requires emergency measures to minimize immediate dangers to surrounding populations, the environment, and private property. Emergency response techniques often include spill stabilization procedures. Some hazardous chemical spill incidents warrant complete chemical cleanup, whereas other spills may be safely contained by temporary or partial spill treatment. Stabilization procedures should be used with caution, however, taking into consideration all potential problem areas such as emergency response personnel safety and logistical requirements.

This chapter discusses the following eight techniques for spill stabilization:

- Mechanical dispersion
- Chemical dispersion
- Water spray knockdown
- Dilution
- Burial
- Neutralization
- Gelling agents
- Sorbents

14.2 MECHANICAL DISPERSION

1. **Objective**—To rapidly disperse spilled materials below harmful concentrations into the air or water a short distance from spill source; compressed air, fans, blowers, water streams, propellers, or rotating paddle wheels may be used.

2. **Applications**—Because air and water pollution result, this technique should be used only when other means of control are not available; may find better applications for reduction of vapors below flammable rather than toxic limits (toxic vapors are harmful at low concentrations).
3. **Limitations**—For large releases, the volume of air or amount of water required becomes limiting; large-capacity equipment may not be readily available; insufficient dispersal could cause more rapid spread of harmful vapors; personnel must get relatively close to spill to set up dispersal systems.
4. **Personnel, Equipment, Materials Required**—Personnel do not need special qualifications; high-volume blowers or fans, air boats, multiple fire streams, and large propeller-driven vessels may be employed.
5. **Spill Size**—Releases small enough to permit equipment to successfully disperse chemicals below harmful concentrations.
6. **Potential Problems**—Air or water contamination.

14.3 CHEMICAL DISPERSION

1. **Objective**—To subject flammable liquid spills to rapid mixing with a surfactant and water, effectively reducing vapor levels below the flammable limit; to assist in biodegradation, dissolution, oxidation, and other natural breakdown processes by increasing spilled material surface area.
2. **Applications**—Effective for reducing flammable chemical vapors from land or water spills; some dispersants may be useful for extinguishing flammable liquid fires.
3. **Limitations**—To be used only in the event of imminent fire danger; otherwise, approval by federal and state on-scene coordinator is required; overapplication may be environmentally harmful.
4. **Personnel, Equipment, Materials Required**—No special personnel training required; backpack pressure sprayers, fire-fighting systems with foam eductors, special dispersant proportioning pumps and spray systems, and commercial aircraft with special spray systems may be used depending on the spill size.
5. **Spill Size**—Not limited by size; logistics of application and dispersant supplies available may limit use on a large scale.
6. **Potential Problems**—May cause increased environmental damage if overapplied; distributes spill throughout water column.

14.4 WATER SPRAY KNOCKDOWN

1. **Objective**—To reduce the concentration of hazardous vapors within a very short distance of a spill or leak source by solubilizing vapors in water.
2. **Applications**—Most effective in removing water-soluble vapors from air; can be used to enhance air dispersion of insoluble vapors as well.
3. **Limitations**—Most effective if application started quickly; large volumes of water and equipment for applying water may be needed when treating large-volume releases; personnel may have to get close to release source to set up water spray; wind may cause problems in water application; runoff from water-soluble releases must be contained.
4. **Personnel, Equipment, Materials Required**—No special personnel training required; adequate water supply and spray equipment needed.
5. **Spill Size**—Limited only by availability of sufficient water and spray equipment.
6. **Potential Problems**—Soil and water contamination may result from unconfined runoff.

14.5 DILUTION

1. **Objective**—To decrease hazardous material concentrations and toxic properties by mixing with water.
2. **Applications**—Useful for retarding or stopping spread of hazardous spills when recovery and removal cannot be accomplished in a timely manner.

14.6 BURIAL

1. **Limitations**—Large amounts of burial material must be readily available; natural forces (wind and rainfall) can expose buried material; technique requires that workers be in close proximity to the spill.
2. **Personnel, Equipment, Materials Required**—Skilled equipment operators needed; depending on spill size, heavy or light equipment necessary for burial; commercial materials such as gypsum, diatomaceous earth, crushed limestone, and various types of clays may be used.
3. **Spill Size**—Almost any volume.
4. **Potential Problems**—Final volume of spilled material may be larger and therefore more difficult to subsequently remove.

14.7 NEUTRALIZATION

1. **Objective**—To apply chemicals that reduce or eliminate the reactive properties of acids or bases.
2. **Applications**—Useful on contained spills of acids or bases on land; useful on spills in quiescent waters such as lakes, rivers, lagoons, harbors, and creeks.
3. **Limitations**—Unconfined spills in moving waterways are difficult to neutralize because of limited application methods; improper application may carry chemical reaction beyond neutralization and produce harmful effects; application of neutralizing agents evenly over spill area is difficult; large amounts of neutralizing agent may be needed; neutralizing chemicals may not be readily available.
4. **Personnel, Equipment, Materials Required**—Personnel must be protected from the potential harmful effects of the neutralizing agent and spilled material; no special training other than use of personal protective equipment required; pH meter, neutralizing agent, shovels, cans, and bags may be used for small spills; pumps, blowers, or other spreading devices are needed for larger spills.
5. **Spill Size**—Most effective on spills of less than 500 gal; larger spills require more planning and application equipment.
6. **Potential Problems**—Overapplication; exposure of personnel to hazardous properties of neutralizing agents; protection of personnel; heat and splattering of spilled material during neutralization reaction; neutralized chemicals may increase volume of final material that must be removed and treated.

14.8 GELLING AGENTS

1. **Objective**—To immobilize spills and thereby reduce vaporization and percolation into the soil.
2. **Applications**—Liquid spills on land; contained and isolated contaminated water; to enhance effectiveness of containment barriers (i.e., gelling a spill to prevent its escape under a containment boom).
3. **Limitations**—High winds hamper application; not readily available in bulk; very costly.
4. **Personnel, Equipment, Materials Required**—No special training for personnel; sufficient gelling agent and compatible gelling agent for spilled

material; application equipment, typically shovels, blowers, cans, and buckets.

5. **Spill Size**—Small spills less than 500 gal.
6. **Potential Problems**—May prevent final spill disposal through pumping and require cleanup by excavation and bulk removal.

14.9 SORBENTS

1. **Objective**—To immobilize and remove spills through absorption or adsorption by sorbent materials.
2. **Applications**—Useful for spills on land; spills on water may be treated if sorbent material is hydrophobic or if a water-soluble chemical is present in very high concentrations.
3. **Limitations**—Sorbents must be compatible with material being treated; treated material must be recovered; sorbents must be applied manually; sorbents may not be readily available in large supply.
4. **Personnel, Equipment, Materials Required**—No special training required for personnel; sufficient sorbent material for spill volume; containers for spent sorbents; pitchforks, rakes, and shovels needed for sorbent dispersal.
5. **Spill Size**—Cost effective for small spills or removal of material that cannot readily or cost effectively be picked up mechanically.
6. **Potential Problems**—Treated material must be recovered; sorbents take on the hazardous properties of the material treated; material needing final disposal is of greater volume; can result in navigational hazards.

Preparing Hazardous Waste for Transportation: DOT Hazardous Waste Transportation Requirements

15.1 INTRODUCTION

The U.S. Environmental Protection Agency (EPA) has adopted by reference the U.S. Department of Transportation (DOT) regulations found in 49 CFR 172, 173, 178, and 179 regarding the transportation of hazardous waste. These relate to packaging, labeling, marking, and placarding. A generator should consult with the transporter as well as the transport, storage, and disposal (TSD) facility prior to preparing hazardous waste for transport, to ensure that the waste is properly packaged, labeled, marked, and placarded.

15.2 DOT REQUIREMENTS

A complete discussion of hazardous materials transportation regulations is beyond the scope of this book and constitutes a book of equal length on just that topic. However, all corporate hazardous waste management programs need to include a systematic means of shipping waste in full compliance with the DOT hazardous materials regulations (49 CFR 100–177). These regulations are voluminous and complex; there is, however, a step-by-step process that can aid companies in meeting their responsibilities to properly describe, classify, package, mark, and label waste materials being offered for transportation. The DOT hazardous waste transportation requirements are summarized below in eight steps.

15.2.1 STEP 1: Classify Material According to DOT Hazard Classes

Any material that will be transported is considered a hazardous material subject to the requirements of 49 CFR if it meets the definition of one or more DOT hazard classes. Following is a list of these hazard classes ordered according to their degree of hazard. The 49 CFR section number where the complete definition can be found is listed in parentheses. The complete definition should be understood before a material is assigned to a hazard class.

- **Radioactive Material (49 CFR 173.389)**—Any material, or combination of materials, that spontaneously gives off ionizing radiation.
- **Poison A (49 CFR 173.326)**—Extremely dangerous poisons. A very small amount of the gas, or vapor of the liquid, mixed with air, is dangerous to life.
- **Compressed Gas–Flammable or Nonflammable (49 CFR 173.300)**—Any material having in the container an absolute pressure exceeding 40 psi at 70°F.
- **Flammable Liquid (49 CFR 173.115a)**—A liquid having a flashpoint below 100°F (140°F).
- **Oxidizer (49 CFR 173.151)**—A substance that yields oxygen readily and accelerates the combustion of organic matter.
- **Flammable Solid (49 CFR 173.240)**—Any solid material that can cause fires through friction or retained heat from manufacturing or processing.
- **Corrosive Material (49 CFR 173.240)**—Any liquid or solid that causes visible destruction or irreversible damage to human skin tissue. Also, a liquid that has a severe corrosion rate on steel.
- **Poison B (49 CFR 173.343)**—Materials so toxic to man that they are a hazard to health during transportation.
- **Irritating Material (49 CFR 173.381)**—Substances that, upon contact with fire or air, give off dangerous or intensely irritating fumes.

- **Combustible Liquid (49 CFR 173.115b)**—Any liquid having a flashpoint at or above 100°F and below 200°F.
- **ORM-B (49 CFR 173.500b2)**—A material the leakage of which could cause significant damage to the vehicle transporting it.
- **ORM-A (49 CFR 173.500b1)**—A material that has an anesthetic, irritating, noxious, toxic, or other similar property.
- **ORM-E (49 CFR 173.500b5)**—A material that is not included in any other hazard class but is a hazardous waste or a hazardous substance.
- **Hazardous Waste (49 CFR 171.8)**—Any material that is subject to the hazardous waste manifest requirements of the U.S. EPA specified in 40 CFR 262.
- **Hazardous Substance (49 CFR 171.8)**—A material (including its mixtures or solutions) that is identified in the CERCLA list of hazardous substances (appendix to the "Hazardous Materials Table" 49 CFR 172.101) when the quantity of the material transported in one package equals or exceeds the reportable quantity (RQ). *Note*: Explosives, organic peroxides, etiologic agents, and ORM-Cs are not included above. They must be classified as required by the regulations.

15.2.2 STEP 2: Select a U.S. DOT Shipping Name for the Material from the Hazardous Materials Table (49 CFR 172.101)

The shipping name selected from the table must be listed with the hazard class identified in Step 1. The regulations include a section titled "Purpose and Use of the Hazardous Materials Table," which explains how to select the proper shipping name for specific materials.

Effective 7/1/87, there are only two proper shipping names available for materials classified as ORM-E: hazardous waste liquid (or solid) N.O.S. and hazardous substance liquid (or solid) N.O.S. The proper shipping name for a material that is a hazardous waste must be preceded by the word "waste" unless it is already included in the name.

15.2.3 STEP 3: Identify Any Additional Description Requirements that May Apply (49 CFR 172.203)

Some hazardous materials require additional information to be included with the proper shipping name. These requirements are detailed in 49 CFR 173.203. Two examples are as follows:

- **Poisonous Materials**—If the name of the compound that causes the material to meet the definition of a poison is not included in the shipping name, it must be added as an additional description.
- **Hazardous Substances**—If the proper shipping name for a material that meets the definition of a hazardous substance (see Step 1) does not include the name(s) of the constituent(s) making it a hazardous substance, the names must be shown in parentheses after the basic description. For materials that are hazardous wastes and assigned a "D", "F", or "K" waste number by the U.S. EPA, the waste code number(s) may be used to meet the additional description in lieu of the name(s) of the constituents.

15.2.4 STEP 4: Packaging the Material as Required

DOT regulations for packaging are found in two different parts of 49 CFR. The "Standard Requirements for All Packages" (49 CFR 173.24) describe basic integrity criteria that any package must meet. In addition to standard requirements, there are detailed requirements for each hazardous material. The section that contains these detailed requirements, as well as the DOT specification packages authorized (if applicable), is noted in the hazardous materials table (49 CFR 172.101), column 5b. Once a proper shipping name is selected, the material must be packaged as required in the section referenced in column 5b for that proper shipping name.

15.2.5 STEP 5: Apply the Required DOT Hazard Warning Labels

The required hazard warning label(s) for each proper shipping name listed in the hazardous material table is indicated in column 4 of the table. There may be additional labels required and not referenced in the hazardous material table. These requirements are detailed in 49 CFR 172.402. Standards for label placement and specification are found in 49 CFR 172.400–172.450.

15.2.6 STEP 6: Mark the Package

DOT requires that each package having a rated capacity of 110 gal or less be marked with the proper U.S. DOT shipping name (preceded by the word "waste" if the material is a hazardous waste), U.N. or N.A. identification number, the letters "RQ" if the material is a reportable quantity of a hazardous substance, and the name(s) of the constituent(s) that makes the material a hazardous substance [if the material is a hazardous waste, the "D", "F", or "K" waste number may be used in lieu of the chemical constituent(s)].

There are numerous other marking requirements that may apply for certain materials. These are found in 49 CFR 172.300–172.338.

15.2.7 STEP 7: Describe the Hazardous Material on a Properly Prepared Shipping Paper (Waste Manifest)

DOT requires that all hazardous materials be accompanied by a properly prepared shipping paper. When the material is a hazardous waste, the uniform hazardous waste manifest must be completed.

The 20 entries on a waste manifest are specified and detailed. 49 CFR 172.200–172.205 and 40 CFR 262.20 detail how these documents are to be completed.

15.2.8 STEP 8: Offer the Hazardous Material for Shipment

DOT requires that certain incompatible materials be separated during transportation. There are additional requirements for bracing containers, placarding vehicles, reporting incidents, routing, and training which must also be followed. No summary or guidelines can fully replace a good working knowledge of the hazardous materials transportation regulations. Shippers (or generators) must ensure that they understand and follow the requirements.

Incident Analysis and Initial Response

16.1 INTRODUCTION

Emergency response operations are made up of three interacting elements:

- Recognition or identification of the substances involved
- Incident evaluation/assessment with respect to hazardous substance behavior and risks posed to people, property, and the environment
- Control measures designed to prevent or reduce existing or potential hazards

These elements are interactive and do not necessarily need to be sequentially implemented. Responding units arriving at a scene to restrict access, for example, may do so without previously identifying spilled material and assessing potential hazards. Other situations may require suspending all response actions until:

- The hazardous substances at the scene are identified, and
- An evaluation of the behavior and toxic properties of these substances has been made

For an effective response operation to be developed, there must be mutual interaction and feedback from the outset among different emergency response elements. The following sections discuss in greater detail each element of the incident analysis and initial response.

16.2 HAZARD RECOGNITION AND IDENTIFICATION

Throughout the initial phases of an emergency response operation, high priority is placed on first identifying which hazardous substances are released at the site as well as their expected behavior in the environment. With this information in hand, informed response strategies can be immediately developed to begin proper management of the incident. Emergency situations in which hazardous substances leak from known sources, such as storage tanks or pipelines, afford straightforward hazard identification. Chemical characterization at hazardous waste sites or highway accidents often proves far more difficult.

The location of an incident within an industrial facility or residential community provides useful insight regarding the potential hazardous substances that may be involved. Facility maps and good community preplanning programs are valuable assets to responders during this phase of the process. Shipping documents (including waybills), consist reports, and shipping papers are also helpful for hazardous chemical identification. Other sources of initial information on the substances at an incident include the following:

- Container placards and labels
- Stenciled container signs
- Tank car or locomotive numbers
- Container shapes

Potential physical clues revealing toxic substances involved in a chemical spill include the following:

- Visible vapor clouds
- Visible flames or frost formation
- Dead animals or fish in the incident area
- Presence of dead or discolored vegetation
- Hissing, venting, or whistling sounds
- Unusual odors

Monitoring instruments can be used at this stage of the response to:

- Detect flammable vapors and low oxygen concentrations
- Detect the presence of very low concentrations of highly toxic organic vapors

Colorimetric indicator tubes also aid verification of the presence of specific inorganic vapors or gases.

16.3 HAZARD EVALUATION AND ASSESSMENT

Initial hazard evaluation and assessment begins immediately upon notification of an incident in progress. The time lapse between notification and arrival at the scene provides responders the opportunity to analyze preliminary incident information concerning:

- The nature of the incident
- Site-specific or modifying conditions
- Potential for hazardous substance exposures
- Immediately available hazardous substance control resources

Once at the scene, the first responder priority should be to estimate the potential harm of the incident. This evaluation generally involves a number of stages:

- Assessing the extent of chemical dispersal
- Determining the hazardous properties of the material present
- Identifying the type of containers involved and their present location
- Noting critical incident conditions including incident location, time, and prevailing weather conditions
- Identifying surrounding populations, property, and ecologically sensitive environments (e.g., lakes and streams) within the affected area

The type of containers involved requires special note. The exact identity and properties of the hazardous substance containers affect the volume of hazardous substance released and the potential environmental exposure pathways. Containers differ in their physical properties and their inherent abilities to withstand stress as from fire or over-pressure conditions. The physical and chemical properties of the hazardous substances in these containers then define the nature and degree of harm present.

The second priority in the assessment phase of response is to determine how and whether response crew involvement can favorably affect an incident. Attempting a rescue, leak repair, or other action without adequate thought to personal protective equipment (PPE) can create more incident problems by exposing more people to incident hazards. In many cases, the nature of the problem as well as the responder resources available will define the most appropriate preventive and corrective actions to be taken. Often, however, emergency response personnel may find that a "no action" course is the best course.

A number of factors may modify the incident or problem situation:

- Incident location (e.g., remote versus highly populated areas)
- Time of day of the incident
- Climatic conditions

In addition, the remoteness of the site, limits on site access, and site proximity to water supplies and other environmentally sensitive areas also affect the choices available to on-site responders.

Although mitigation of the incident is important, protection of both civilian and emergency response personnel must be the principal concern guiding emergency response actions. The following populations are at particular risk during an incident:

- Nursing homes
- Schools
- Hospitals
- Stadiums
- Auditoriums

These facilities should be promptly identified and considered in the planning of response actions.

As indicated above, communities at risk from hazardous chemical exposure also include environmentally sensitive areas, property, and equipment. Naturally, property and equipment receive the lowest priority in an emergency response situation. Effects on the environment, as summarized in the discussion on chemical toxicity, can be short term or long lasting; from a risk management standpoint, elimination of long-term chronic environmental problems, such as pollution of rivers, streams, and soils, should be considered when determining how to respond to an incident.

Once the dimensions of a chemical or hazardous substance release have been assessed, the resources available for emergency response use will be identified. Internal resources include the following:

- Available personnel and their expertise and training
- PPE
- Communications
- Specialized response equipment
- Control agents (such as foam)
- Supportive resources such as aerial surveillance, dispersion modeling, and public affairs

There are also a number of industry-standard resources that can be used at this stage of the response:

- Special emergency response groups (e.g., *Chlororep*, *Chemtrec*)
- Local emergency services (e.g., police department, sheriff's department, emergency management agency, fire department)
- Mutual aid groups and medical care groups

16.4 REFERENCES

1. U.S. DOT, U.S. Coast Guard. *CHRIS—Hazard Assessment Handbook.* CG-446-3. Washington, DC, 1974.
2. U.S. EPA. *Hazardous Materials Incident Response Training Program,* 1983.
3. Canadian Chemical Producers Association. *A Disciplined Approach to Emergency.* Esso Chemicals, 1982.

17

Incident Command System

17.1 INTRODUCTION

The Incident Command System (ICS) is a management protocol to address all critical issues during the response to an emergency incident. The ICS consists of personnel, facilities, equipment, communications, and procedures all operating within a common organizational structure to gain control of an incident. With ICS in place, a complete organization is available that includes the following branches:

- Command
- Operations
- Planning
- Logistics
- Finance
- Command support functions: safety, liaison, staging, and information

The ICS is designed to allow for multiple agency participation, including industrial, federal, state, and local emergency agencies. ICS terminology is therefore designed for acceptance and implementation by all levels of industry and government. Because the ICS is a basic everyday operating system for all incidents, transition of the system to large operations requires a minimum of adjustment. For example, the ICS initially bestows management of all major incident functions on one or a few persons; as the incident grows in size and/or complexity, management of all activities can be assigned to additional individuals to maintain the appropriate level of control and efficiency. The basic organizational structure is applicable to small incidents, hazardous materials spills, major industrial fire incidents, and incidents not directly involving fires and spills, such as hurricanes, floods, rescues, or terrorist events. The organizational structure of the system is able to adapt to any incident to which personnel would be expected to respond.

Organization of the ICS begins when an incident occurs and continues until incident management and operations are no longer necessary. ICS organization develops as the incident progresses, depending on the requirements of the situation. Positions within the ICS will be activated only when required by the incident. Personnel are responsible for all activities under their organizational section for the duration of the incident.

17.2 PRINCIPAL COMPONENTS OF THE INCIDENT COMMAND SYSTEM

A number of components provide the basis for an effective ICS.

17.2.1 Terminology

Common terminology exists for each of the following elements:

- **Position Titles**—A standard set of position titles and major functions and functional units has been established for the ICS. Terminology for the elements is standard and consistent (e.g., operations chief, divisions, etc.).
- **Resource Titles and Specifications**—Common names and specifications have been established for all resources used within the ICS.
- **Facilities**—Common identifiers are used for those facilities in and around the incident area that will be used during the course of the incident, such as the command post and staging areas.

17.2.2 Flexible Organizational Structure

The ICS structure builds from the top down, with responsibility and performance placed initially with the Command Section. As the need exists, separate sections can be developed and added (Operations, Planning, Logistics, and Finance). The organization established for any given incident will be based on the needs of the incident. If one person can simultaneously manage all major functional areas, no other positions are required.

17.2.3 Integrated Communications Procedures

Communications at the incident are managed through the use of a common communications plan and are coordinated through the Incident Command. All communications at an incident should be in common English. No special codes should be employed, and all communications should be confined to essential messages. Face-to-face communications are the most effective method of information exchange during an emergency situation and should be used where practical; face-to-face communications not only minimize the potential for incorrect communication of essential information, but also reduce radio congestion.

As an emergency response or hazardous materials management crisis escalates, multiple radio frequencies can be used simultaneously as follows:

- **Dispatch**—Used to issue standby and assistance alerts, all clear, and to request resource response.
- **Command**—Used to link Command with key command staff members, tactical division officers, and a dispatch center.
- **Tactical**—Tactical frequencies are established when several divisions or groups of resources have a need for radio coordination.
- **Support**—Support frequencies may be established to handle requests for support, logistics coordination, and other nontactical needs.

17.2.4 Unified Command Structure

Proper selection of participants to work within a command structure will depend on:

- The location of the incident—which companies, cities, and other jurisdictions are involved
- The kind of incident—which agencies of the involved jurisdiction(s) are required

A unified command structure should consist of a key responsible official serving as the Incident Commander selected by the organization requesting assistance. The Incident Commander can also invite the counsel of responding individuals or agencies having special expertise or capabilities.

17.2.5 Action Plan

Every incident needs an action plan. For small incidents of short duration, the plan often is not written. The Incident Commander is charged with the establishment of this plan and making the necessary strategic determinations for the incident based on the situation and its required actions.

Safety factors, as well as sound management planning, will influence and dictate the span of control responsibilities. Within the ICS, the span of control of any individual with emergency management responsibility should range from three to seven persons.

17.2.6 Incident Facilities

There are several kinds and types of facilities that can be established in and around the incident area. The needs determination for incident facilities and their locations during an emergency situation will be based on the requirements of the incident and the direction of Command. The following facilities are defined for possible use with the ICS:

- **Incident Command Post**—Designated as "Command," the command post is the location from which all incident operations are directed. The Incident Commander is based at the command post. There is traditionally only one command post for the incident.
- **Incident Base**—The incident base is where primary support activities are performed. The base houses all equipment and personnel support operations until these resources are assigned to a staging area.
- **Emergency Operations Center (EOC)**—The EOC is a location away from the emergency situation having access to the senior facility management and radio/telephone communications.
- **Staging Areas**—The staging area is established for temporary location of available resources on 3-min notice. The staging area will be established by Command to control resources not immediately assigned.

17.2.7 Comprehensive Resource Management

Equipment and personnel involved in an incident must be used to their full potential to bring an incident to a quick, safe resolution. One of the most productive actions by Command is to first group personnel and equipment into strike teams or task forces; this action maximizes both resource use and control of a large number of single resources while reducing the communications load. One important, but often overlooked, issue is to ensure that the status of resources be kept current and accurate.

17.3 ORGANIZATION AND OPERATIONS

The ICS organization has five major functional areas:

- Command

- Operations
- Planning
- Logistics
- Finance

17.3.1 Command

17.3.1.1 Introduction

Command is responsible for overall management of all incident activities, including certain staff functions such as:

- Safety Officer
- Government Liaison Officer
- Information Officer

The transition from initial command structure developed upon arrival at the site to deployment of major incident organization will be evolutionary; positions are most often filled as the need for specific tasks is identified. Examples of incident organization structure listing all positions reasonably required for large-scale incidents are provided at the end of this chapter (see Figures 17.1 to 17.3).

The Incident Commander develops a series of incident objectives upon which all subsequent action planning by the Incident Commander as well as subordinate staff are based. The Incident Commander must always remain knowledgeable of the incident action plan and should approve all requests for procuring primary resources. Often one or several governmental or regulatory agencies will have some jurisdiction at an incident due to:

- The nature of the incident, or
- The kinds of resources required

For example, almost any accident involving marine release of fuels or an uncontrolled release of natural gas will involve governmental or regulatory agencies with overlapping responsibilities and jurisdictions. In these situations, the Incident Commander will ensure that all of these organizations are coordinately involved as much as possible as:

- Important response and resource decisions are made
- Response strategies are developed, and
- Action plans are implemented

17.3.1.2 Command

"Command" is the appropriate identifier used by on-scene personnel involved in the implementation of response strategies and action plans. Command is traditionally established by those who arrive first on the scene with the necessary position to direct the actions of all incoming responders. As soon as possible, an initial incident announcement should be distributed providing:

- An initial report on the emergency situation, and
- Identification and location of Command

It is also important to distribute a brief description of the emergency—known as the size-up—to give those responding to the situation a chance to prepare mentally for the conditions present.

As senior officials arrive, Command may be transferred to more senior personnel. As personnel are given incident assignments, Command should announce such assignments in regular briefings. For example, "George is now the Staging Officer and staging is being established at the corner of 5th and E Streets. Judy is now Press Officer and the press office is being established in the police station on the corner of 5th and G Streets." At the conclusion of the incident, Command will be terminated by the Incident Commander, who provides this information to all parties affected.

Command should determine when to divide the organizational responsibility for each incident by creating sections for each of the following functions:

- Operations
- Planning
- Logistics
- Finance

As new sections are created, section chiefs assume immediate direction and control. For instance, an Operations Section Chief, typically the first section created during an incident, manages tactical implementation of the action plan developed by Command.

The following factors typically encourage Command to divide organization of the incident into separate sections:

- When Command can mentally forecast a situation that will eventually involve a number of responders beyond his or her control capability. A good rule of thumb for creating new sections is whenever there are more than five persons or teams reporting to Command.

- When Command can no longer effectively observe all responder activities because of physical barriers (e.g., responders are inside a building or are on the opposite side of a tank).
- When situations arise where close coordination of personnel is required, such as a hazardous materials response or elevated structure rescue.

17.3.1.3 Safety Officer

The Safety Officer's function at the incident is to assess hazardous and unsafe situations and develop measures for ensuring personnel safety. The Safety Officer will correct unsafe acts or conditions through the regular line of authority, although he or she may exercise emergency authority to stop and/or prevent unsafe acts or conditions when immediate action is required.

Standard operating procedures (SOPs) should be established to better assist Command. The Safety Officer must ensure that health and safety of personnel at incidents remains the highest priority. SOPs for protective clothing, establishment of hazardous work zones, evacuation routes, special radio procedures, and so forth should be an integral part of a facility's initial plan or pre-plan.

17.3.1.4 Government Liaison Officer

The Liaison Officer's function is to be a designated point of contact for representatives from other agencies. Representatives from assisting agencies coordinate all of their efforts with Command through the Liaison Officer. It should be noted here that it is critical that agency representatives assigned to an incident have the proper authority to speak for their agency during emergency response situations.

17.3.1.5 Information Officer

The Information Officer obtains briefings from the Incident Commander and provides information to the press and media about the incident. One of the first tasks of the Information Officer during an emergency is to establish an area for the press away from the command post to ensure that press activities do not interfere with or distract the Incident Commander's ability to properly manage emergency activities. Unsafe areas and areas with ongoing operations should be noted, and the press should be clearly instructed not to enter these areas while the emergency response is in progress. The Information Officer should also be aware of what information is not to be released to the press until a complete investigation has taken place. Examples of sensitive information often not made public until after an incident has been properly managed include:

- The names of accident victims
- The cause of the incident

17.3.1.6 Staging Officer

A Staging Officer, designated by the Incident Commander, will establish and coordinate the staging area to temporarily locate resources that are available for assignment during the emergency. In some cases, the Staging Officer may report to the Incident Commander or to the Operations Section Chief, depending on the particular circumstances of the incident and the Command in place at the scene. During operation of the staging area, the Staging Officer can request logistical support (such as food, fuel, etc.) from Logistics; Command may establish, move, or discontinue the use of the staging area as the needs of the emergency change.

The Staging Officer is responsible for the following:

- All staged equipment is positioned in an appropriate manner
- A log is maintained accurately documenting:
  - The resources available to emergency components
  - All specialized equipment
- Communication with Command is maintained to monitor what level(s) of equipment staging should be maintained during response. Requests for resources in Staging should be made to Logistics.

17.3.2 Operations

17.3.2.1 Introduction

The Operations Section is managed by the Operations Chief, who is responsible for direct management of all incident tactical activities. In any incident, the individual resources assigned to the response are at the disposal of the individual who has overall responsibility—namely Command. As the incident grows in size and complexity, Command may designate an Operations Chief to assume tactical direction of resources.

17.3.2.2 Operations Section Organization

17.3.2.2.1 Divisions and Groups

Divisions and groups are established at an incident when the number of resources (in the form of single increments, task forces, or strike teams) exceeds the span of control of Command or the Operations Chief.

Divisions are established to divide an incident into *geographical* areas of operation. The best use of divisions requires that they be divided:

- Within an area according to natural separations of terrain, geography, and fuel, and
- Where resources can be effectively managed under span-of control guidelines

Groups are established to divide an incident into *functional* areas of operation. Groups can best be used to describe areas of like activity—for example, rescue, spill containment, medical, foam, and decontamination.

Both divisions and groups may be used at a single incident if justified and if proper coordination can be effected. Branches of divisions or groups may also be established at an incident to serve several purposes:

- When the number of divisions and/or groups exceeds the span of control for the Operations Chief
- When the nature of the incident calls for a functional branch structure
- When the incident is multi-jurisdictional and where resources are best managed under the agencies that have normal control over those resources

17.3.2.2.2 Sectors

The term "sector" is not used in the original ICS. However, sectors are used in other command systems to designate the splitting of resources either geographically or functionally.

17.3.2.2.3 Duties of Division/Group/Branch Officers

Officers assigned to an area or function evaluate and report on emergency response and incident conditions to the Operations Chief, if assigned, or to Command. These officers are designated by the area or function that they manage (e.g., North Division, Decontamination Group, Foam Branch, etc.) and are usually responsible for the safety and performance of personnel and resources in their respective areas. Requests for resources and information from these officers are directed to the Operations Chief.

17.3.3 Planning

The Planning Section is responsible for the collection, evaluation, and dissemination of tactical information. Planning maintains information on the current and predicted or forecast situation as well as on the status of resources assigned to the incident. Planning is also responsible for the preparation and documentation of action plans. The Planning Section often has several components and may be staffed by a variety of people:

- Technical specialists to assist in evaluating the situation
- Forecasting specialists to determine the requirements for additional personnel and equipment

The Planning Section is usually managed by a Planning Chief appointed by Command.

17.3.3.1 Resources/Situation Officer

The Resources/Situation Officer is responsible for maintaining the current status of all resources. An accurate resource status management system is required to show the current status and location of all incident resources. This management activity should include accounting for the use and deployment of the following:

- Key supervisory personnel
- Primary resources in use in tactical operations
- Support resources
- Transportation equipment

The Resources/Situation Officer is also responsible for:

- Collecting, processing, and organizing situation information
- Preparing situation summaries
- Developing projections and forecasts of future events related to the incident

17.3.3.2 Demobilization Officer

The Demobilization Officer is responsible for developing an incident demobilization plan. The plan should include specific demobilization instructions for all resources that require demobilization.

17.3.3.3 Technical Specialists

Technical specialists assigned to the Planning Section report to the Planning Chief, Command, or the Operations Chief. The nature of the incident traditionally will dictate the need for, as well as the types of, technical specialists required. The following specialists are often needed at chemical emergency sites:

- Meteorologists
- Environmental engineers
- Chemists
- Industrial hygienists
- Product specialists
- Water system specialists
- Structural engineers
- Training specialists

17.3.4 Logistics

Logistics is responsible for providing all support needs at the incident. The Logistics Chief is appointed by Command in larger or more complex incidents. Logistics carries out a number of important incident functions:

- Arranges ordering of all resources from off-incident locations
- Provides facilities, transportation, supplies, equipment maintenance, fueling, feeding, and communications support

The following positions may be created under the Logistics Section:

- Supply Officer
- Facilities Officer
- Ground Support Officer
- Communications Officer
- Food Officer
- Emergency Incident Dispatch Officer

The Supply Officer is responsible for ordering, receiving, storing, and processing all incident-related resources, personnel, and supplies.

The Facilities Officer establishes, maintains, and closes all facilities used in support of incident operations. The Facilities Officer may establish the command post, incident base, and staging areas, which are often established in areas where there are existing structures that can be employed.

The Ground Support Officer has the following responsibilities:

- Maintenance and repair of primary tactical equipment, vehicles, and mobile ground support equipment
- Fueling of all mobile equipment
- Providing transportation services in support of incident operations

In addition to maintenance and service of all mobile vehicles and equipment, Ground Support will, on major incidents, maintain a transportation pool. The transportation pool will consist of vehicles that can be used to transport personnel. It must also provide Planning personnel with up-to-date information on the status of transportation vehicles, their locations, and capabilities.

The Communications Officer must develop plans to make the most effective use of:

- Incident-assigned communications equipment and facilities
- Installation and testing of all communications equipment
- Distribution and recovery of spare equipment assigned to incident personnel
- Maintenance and on-site repair of communications equipment

Communications in the ICS has a major responsibility for effective communications planning because of the potential for multi-agency use of the system. Communications is especially important in determining required radio nets, establishing inter-agency frequency assignments, and ensuring that maximum use is made of assigned communications capability.

The Food Officer is responsible for determining food and water requirements, food ordering and serving, and general maintenance of the food service area. Food services is an extremely important part of incident operations. Food services must be able to anticipate incident needs both in terms of numbers of personnel to be fed and any special feeding requirements due to the kind or location of the incident. Food services is responsible for supplying the food needs for the entire incident, including all remote locations (e.g., staging areas), as well as supplying food service to Operations personnel unable to leave tactical assignments. This division must interact closely with a number of other divisions:

- Planning—to determine personnel requirements
- Facilities Officer—for fixed feeding locations
- Supply personnel—for food ordering
- Ground Support personnel—for supplying transportation services

Emergency Incident Dispatch Officers maintain a listing of available emergency resources and request those resources as directed by Command.

The Rehabilitation Section is located in a safe area away from the incident, where all incident personnel can obtain refreshments as well as needed rest. Water, electrolyte fluids, and food should be available—particularly for incidents over extended time periods. Personnel in Rehabilitation will not be considered available for service until they have been released by the Rehabilitation Officer to Command.

17.3.5 Finance

A Finance Section is established for incidents that need integrated or substantial financial services. The Finance Chief, when established by Command, determines the need for establishing specific units to cover the following areas:

- Time
- Procurement
- Compensation/claims
- Cost

The Time Officer is responsible for maintaining payroll records of personnel involved in incident operations to ensure proper compensation is provided.

The Procurement Officer manages the purchase of equipment and supplies needed to manage a particular incident. The Procurement Officer must work closely with personnel in the Logistics Section to ensure that ordered materials are received and that proper prices are paid.

Compensation/Claims Officers document and handle all claims for losses occurring as a result of an emergency or incident. They must work closely with other agencies providing incident supplies to the public when personal property has been damaged by incident or response activities.

The Cost Officer is responsible for documenting the total value of losses due to the incident, including product value and property damage.

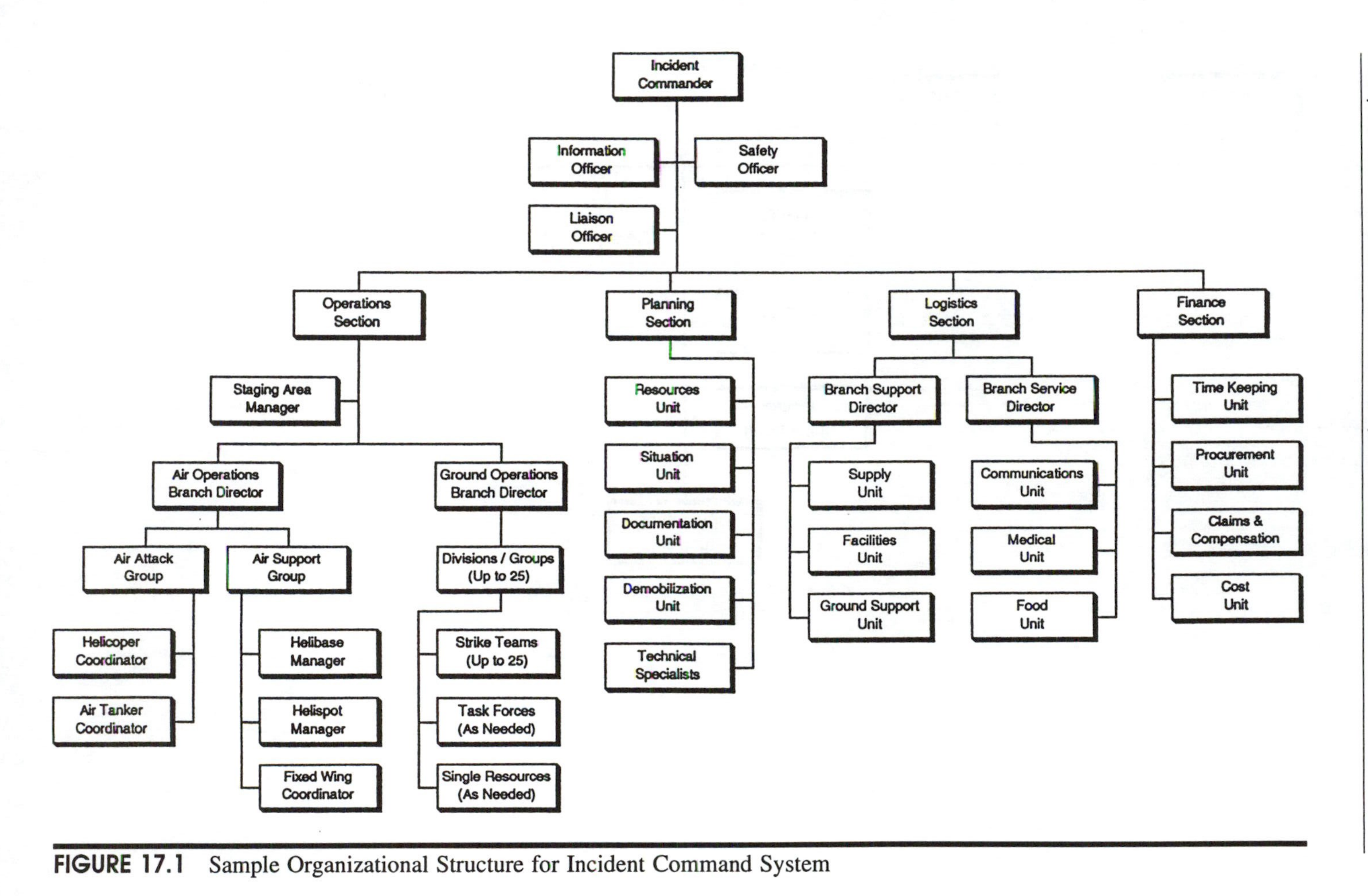

FIGURE 17.1 Sample Organizational Structure for Incident Command System

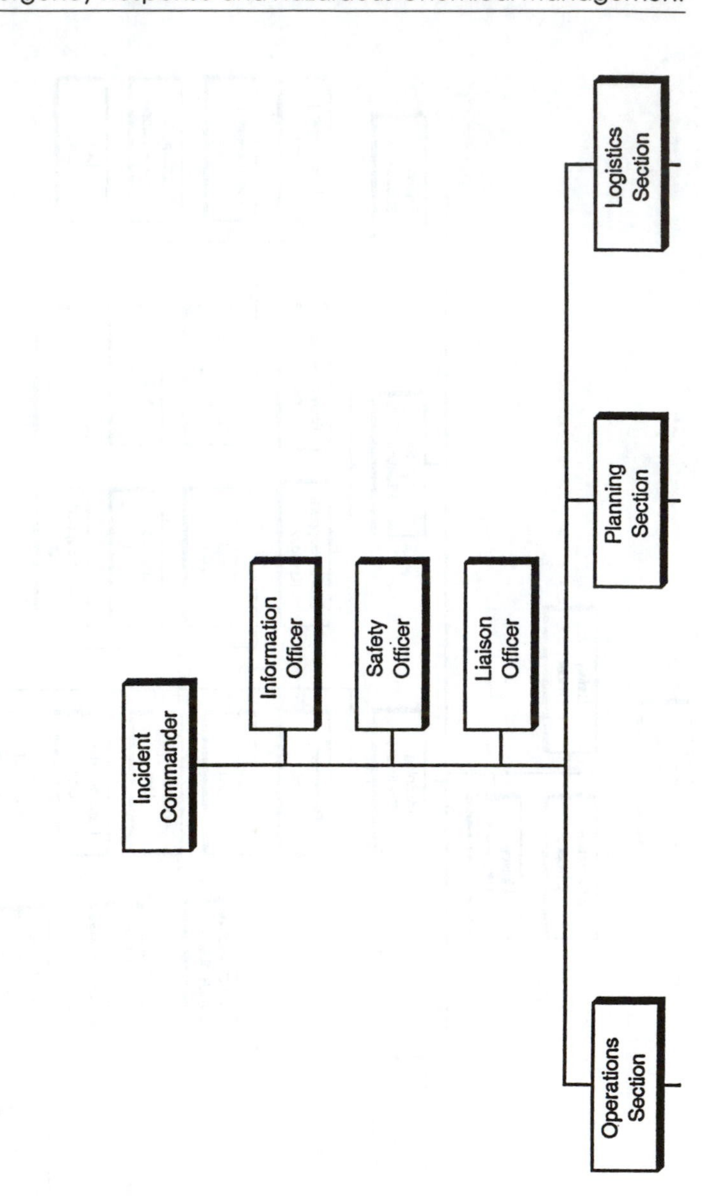
Incident Commander
Information Officer
Safety Officer
Liaison Officer
Operations Section
Planning Section
Logistics Section

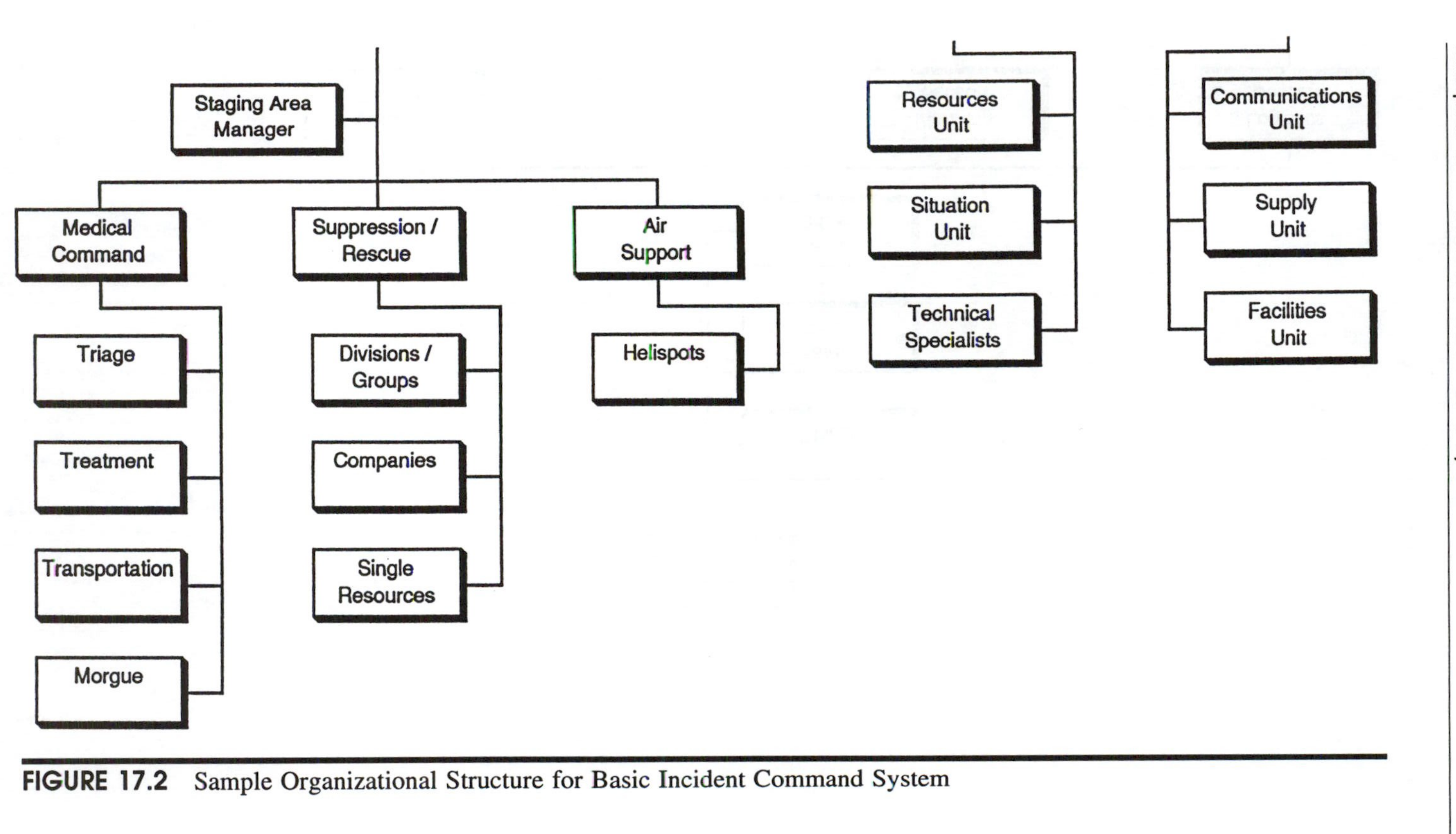

FIGURE 17.2 Sample Organizational Structure for Basic Incident Command System

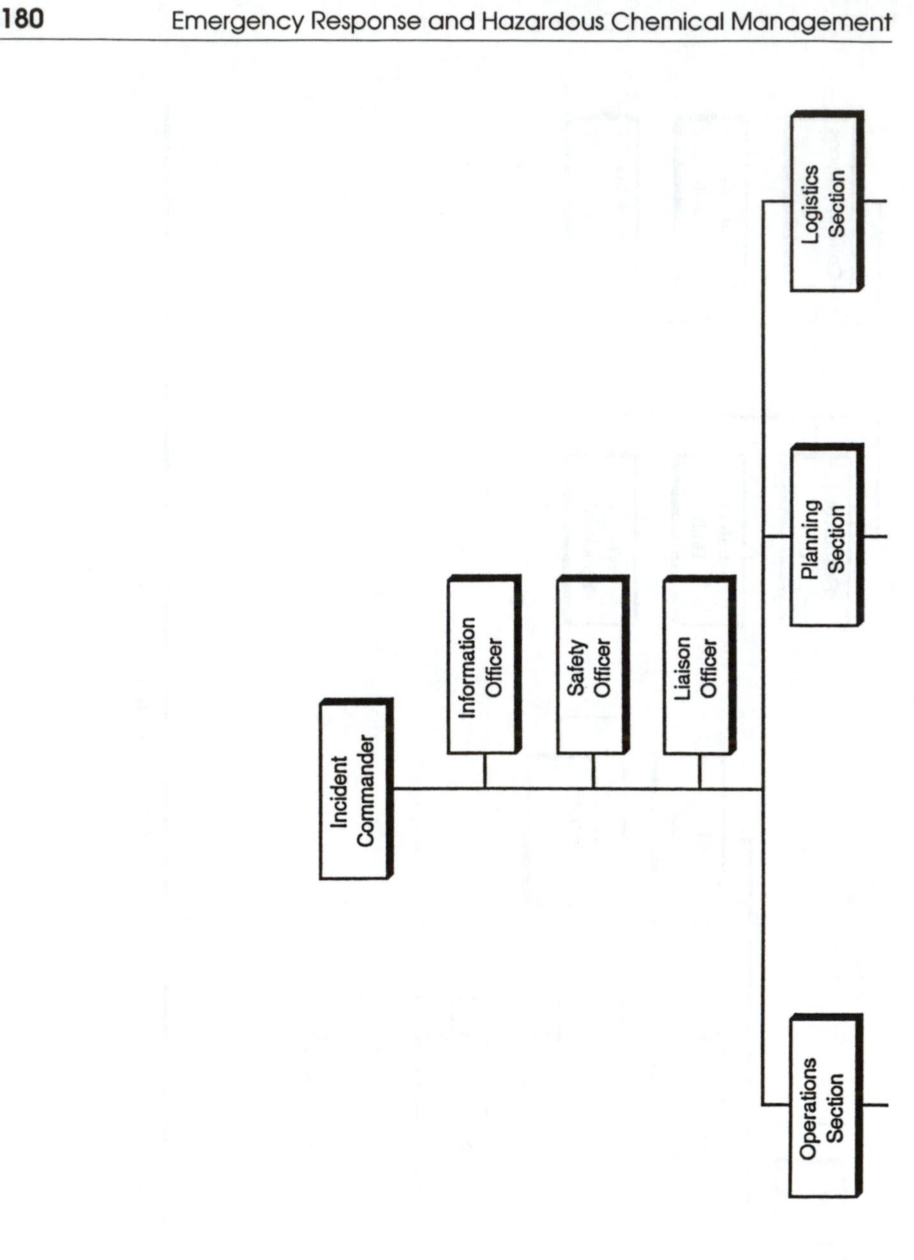

Incident Commander
Information Officer
Safety Officer
Liaison Officer
Operations Section
Planning Section
Logistics Section

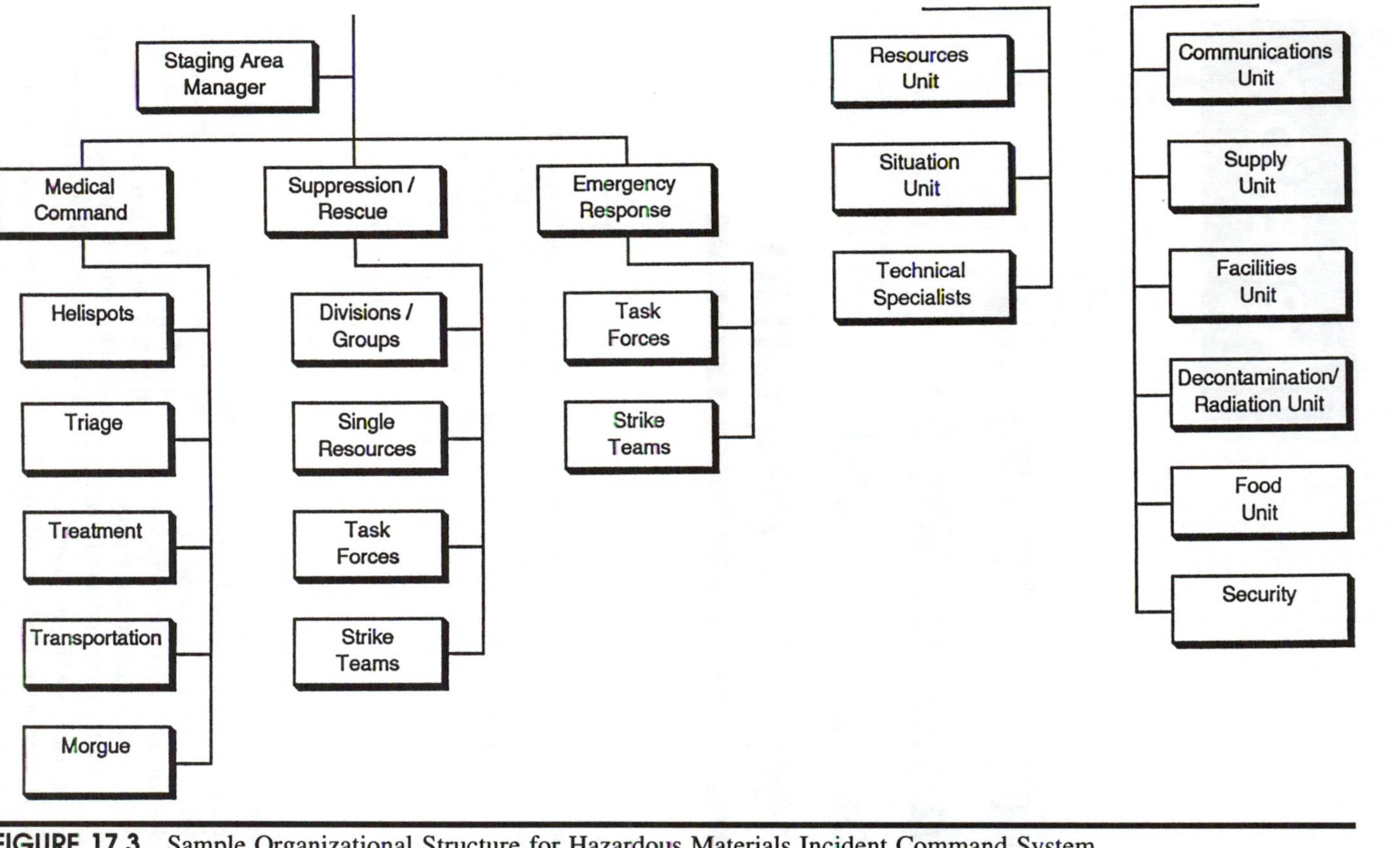

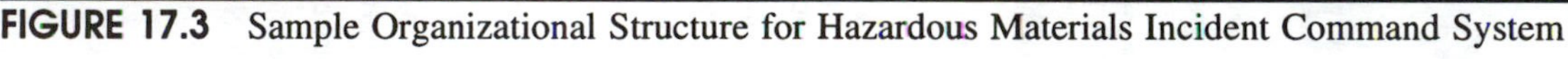

FIGURE 17.3 Sample Organizational Structure for Hazardous Materials Incident Command System

29 CFR 1910.120 Hazardous Waste Operations and Emergency Response

(a) Scope, application, and definitions— (1) Scope. This section covers the following operations, unless the employer can demonstrate that the operation does not involve employee exposure or the reasonable possibility for employee exposure to safety or other health hazards.

(a)(1)(i) Clean-up operations required by a governmental body, whether Federal, state, local or other involving hazardous substances that are conducted at uncontrolled hazardous waste sites (including, but not limited to, the EPA's National Priority Site List (NPL), state priority site lists, sites recommended for the EPA NPL, and initial investigations of government identified sites which are conducted before the presence or absence of hazardous substances has been ascertained);

(a)(1)(ii) Corrective actions involving clean-up operations at sites covered by the Resource Conservation and Recovery Act of 1976 (RCRA) as amended (42 USC 6901 *et seq.*);

(a)(1)(iii) Voluntary clean-up operations at sites recognized by Federal, state, local or other governmental bodies as uncontrolled hazardous waste sites;

(a)(1)(iv) Operations involving hazardous wastes that are conducted at treatment, storage, and disposal (TSD) facilities regulated by 40 CFR Parts 264 and 265 pursuant to RCRA; or by agencies under agreement with USEPA to implement RCRA regulations; and

(a)(1)(v) Emergency response operations for releases of, or substantial threats of releases of, hazardous substances without regard to the location of the hazard.

(a)(2) Application. (i) All requirements of Part 1910 and Part 1926 of Title 29 of the Code of Federal Regulations apply pursuant to their terms of hazardous waste and emergency response operations whether covered by this section or not. If there is a conflict or overlap, the provision more protective of employee safety and health shall apply without regard for 29 CFR 1910.5(c)(1).

(a)(2)(ii) Hazardous substance clean-up operations within the scope of paragraphs (a)(1)(i) through (a)(1)(iii) of this section must comply with all paragraphs of this section except paragraphs (p) and (q).

(a)(2)(iii) Operations within the scope of paragraph (a)(1)(iv) of this section must comply only with the requirements of paragraph (p) of this section.

Notes and Exceptions:

(a) All provisions of paragraph (p) of this section cover any treatment, storage, or disposal (TSD) operation regulated by 40 CFR parts 264 and 265 or by state law authorized under RCRA, and required to have a permit or interim status from EPA pursuant to 40 CFR 270.1 or from a state agency pursuant to RCRA.

(b) Employers who are not required to have a permit or interim status because they are conditionally exempt small quantity generators under 40 CFR 261.5 or are generators who qualify under 40 CFR 262.34 for exemptions from regulation under 40 CFR parts 264, 265 and 270 ("excepted employers") are not covered by paragraphs (p)(1) through (p)(7) of this section. Excepted employers who are required by the EPA or state agency to have their employees engage in emergency response or who direct their employees to engage in emergency response are covered by paragraph (p)(8) of this section, and cannot be exempted by (p)(8)(i) of this section. Excepted employers who are not required to have their employees engage in emergency response, who direct their employees to evacuate in the case of such emergencies and who meet the requirements of paragraph (p)(8)(i) of this section are exempt from the balance of paragraph (p)(8) of this section.

(c) If an area is used primarily for treatment, storage or disposal, any emergency response operations in that area shall comply with paragraph (p)(8) of this section. In other areas not used primarily for treatment, storage, or disposal, any emergency response operations shall comply with paragraph (q) of this section. Compliance with the requirements of paragraph (q) of this section shall be deemed to be in compliance with the requirements of paragraph (p)(8) of this section.

[55 F.R. 14073, April 13, 1990]

(a)(2)(iv) Emergency response operations for release of, or substantial threats of releases of, hazardous substances which are not covered by paragraphs (a)(1)(i) through (a)(1)(iv) of this section must only comply with the requirements of paragraph (q) of this section.

(a)(3) Definitions. "*Buddy system*" means a system of organizing employees into work groups in such a manner that each employee of the work group is designated to be observed by at least one other employee in the work group. The purpose of the buddy system is to provide rapid assistance to employees in the event of an emergency.

"*Clean-up operation*" means an operation where hazardous substances are removed, contained, incinerated, neutralized, stabilized, cleared-up, or in any other manner processed or handled with the ultimate goal of making the site safer for people or the environment.

"*Decontamination*" means the removal of hazardous substances from employees and their equipment to the extent necessary to preclude the occurrence of foreseeable adverse health affects.

"*Emergency response*" or "*responding to emergencies*" means a response effort by employees from outside the immediate release area or by other designated responders (i.e., mutual-aid groups, local fire departments, etc.) to an occurrence which results, or is likely to result, in an uncontrolled release of a hazardous substance. Responses to incidental releases of hazardous substances where the substance can be absorbed, neutralized, or otherwise controlled at the time of release by employees in the immediate release area, or by maintenance personnel, are not considered to be emergency responses within the scope of this standard. Responses to releases of hazardous substances where there is no potential safety or health hazard (i.e., fire, explosion, or chemical exposure) are not considered emergency responses.

"*Facility*" means (A) any building, structure, installation, equipment, pipe or pipeline (including any pipe into a sewer or publicly owned treatment works), well, pit, pond, lagoon, impoundment, ditch, storage container, motor vehicle, rolling stock, or aircraft, or (B) any site or area where a hazardous substance has been deposited, stored, disposed of, or placed, or otherwise came to be located; but does not include any consumer product in consumer use or any water borne vessel.

"*Hazardous materials response (HAZMAT) team*" means an organized group of employees, designated by the employer, who are expected to perform work to handle and control actual or potential leaks or spills of hazardous substances requiring possible close approach to the substance. The team members perform responses to releases or potential releases of hazardous substances for the purpose of control or stabilization of the incident. A HAZMAT team is not a fire brigade nor is a typical fire brigade a HAZMAT team. A HAZMAT team, however, may be a separate component of a fire brigade or fire department.

"*Hazardous substance*" means any substance designated or listed under paragraphs (A) through (D) of this definition, exposure to which results or may result in adverse affects on the health or safety of employees:

(a) Any substance defined under section 101(14) OF CERCLA;

(b) Any biological agent and other disease-causing agent which after release into the environment and upon exposure, ingestion, inhalation, or assimilation into any person, either directly from the environment or indirectly by ingestion through food chains, will or may reasonably be anticipated to cause death, disease, behavioral abnormalities, cancer, genetic mutation, physiological malfunctions (including malfunctions in reproduction) or physical deformations in such persons or their offspring;

[55 F.R. 14073, April 13, 1990]

(c) Any substance listed by the US Department of Transportation as hazardous materials under 49 CFR 172.101 and appendices; and

(d) Hazardous waste as herein defined.

"*Hazardous waste*" means

(a) A waste or combination of wastes as defined in 40 CFR 261.3, or

(b) Those substances defined as hazardous wastes in 49 CFR 171.8.

"*Hazardous waste operation*" means any operation conducted within the scope of this standard.

"*Hazardous waste site*" or "*Site*" means any facility or location within the scope of this standard at which hazardous waste operations take place.

"*Health hazard*" means a chemical, mixture of chemicals or a pathogen for which there is statistically significant evidence based on at least one study conducted in accordance with established scientific principles that acute or chronic health effects may occur in exposed employees. The term "health hazard" includes chemicals which are carcinogens, toxic or highly toxic agents, reproductive toxins, irritants, corrosives, sensitizers, hepatotoxins, nephrotoxins, neurotoxins, agents which act on the hematopoietic system, and agents which damage the lungs, skin, eyes, or mucous membranes. It also includes stress due to temperature extremes. Further definition of the terms used above can be found in Appendix A to 29 CFR 1910.1200.

"*IDLH*" or "*Immediately dangerous to life or health*" means an atmospheric concentration of any toxic, corrosive or asphyxiant substance that poses an immediate threat to life or would cause irreversible or delayed adverse health effects or would interfere with an individual's ability to escape from a dangerous atmosphere.

"*Oxygen deficiency*" means that concentration of oxygen by volume below which atmosphere supplying respiratory protection must be provided. It exists in atmospheres where the percentage of oxygen by volume is less than 19.5 percent oxygen.

"*Permissible exposure limit*" means the exposure, inhalation or dermal permissible exposure limit specified in 29 CFR Part 1910, Subparts G and Z.

"*Published exposure level*" means the exposure limits published in "NIOSH Recommendations for Occupational Health Standards" dated 1986 incorporated by reference, or if none is specified, the exposure limits published in the standards specified by the American Conference of Governmental Industrial Hygienists in their publication "Threshold Limits Values and Biological Exposure Indices for 1987–88" dated 1987 incorporated by reference.

"*Post emergency response*" means that portion of an emergency response performed after the immediate threat of a release has been stabilized or eliminated and clean-up of the site has begun. If post-emergency response is performed by an employer's own employees who were part of the initial emergency response, it is considered to be part of the initial response and not post-emergency response. However, if a group of an employer's own employees, separate from the group providing initial response, performs the clean-up operation, then the separate group of employees would be considered to be performing post-emergency response and subject to paragraph (q)(11) of this section.

"*Qualified person*" means a person with specific training, knowledge and experience in the area for which the person has the responsibility and the authority to control.

"*Site safety and health supervisor (or official)*" means the individual located on a hazardous waste site who is responsible to the employer and has the authority and knowledge necessary to implement the site safety and health plan and verify compliance with applicable safety and health requirements.

"*Small quantity generator*" means a generator of hazardous wastes who in any calendar month generates no more than 1,000 kilograms (2,205 pounds) of hazardous waste in that month.

"*Uncontrolled hazardous waste site*" means an area where an accumulation of hazardous waste creates a threat to the health and safety of individuals or the environment or both. Some sites are found on public lands, such as those created by former municipal, county or state landfills where illegal or poorly managed waste disposal has taken place. Other sites are found on private property, often belonging to generators or former generators of hazardous waste. Examples of such sites include, but are not limited to, surface impoundments, landfills, dumps, and tank or drum farms. Normal operations at TSD sites are not covered by this definition.

(b) Safety and health program—

Note to (b): Safety and health programs developed and implemented to meet other Federal, state, or local regulations are considered acceptable in meeting this requirement if they cover or are modified to cover the topics required in this paragraph. An additional or separate safety and health program is not required by this paragraph.

(b)(1) General. (i) Employers shall develop and implement a written safety and health program for their employees involved in hazardous waste operations. The program shall be designed to identify, evaluate, and control safety and health hazards, and provide for emergency response for hazardous waste operations.

(b)(1)(ii) The written safety and health program shall incorporate the following:

(b)(1)(ii)(a) An organizational structure;

(b)(1)(ii)(b) A comprehensive workplan;

(b)(1)(ii)(c) A site-specific safety and health plan which need not repeat the employer's standard operating procedures required in paragraph (b)(1)(ii)(f) of this section;

(b)(1)(ii)(d) The safety and health training program;

(b)(1)(ii)(e) The medical surveillance program;

(b)(1)(ii)(f) The employer's standard operating procedures for safety and health; and

(b)(1)(ii)(g) Any necessary interface between general program and site specific activities.

(b)(1)(iii) Site excavation. Site excavations created during initial site preparation or during hazardous waste operations shall be shored or sloped as appropriate to prevent accidental collapse in accordance with Subpart P of 29 CFR Part 1926.

(b)(1)(iv) Contractors and subcontractors. An employer who retains contractor and subcontractor services for work in hazardous waste operations shall inform those contractors, subcontractors, or their representatives of the site emergency response procedures and any potential fire, explosion, health, safety or other hazards of the hazardous waste operation that have been identified by the employer, including those identified in the employer's information program.

(b)(1)(v) Program availability. The written safety and health program shall be made available to any contractor or subcontractor or their representative who will be involved with the hazardous waste operation; to employees; to employee designated representatives; to OSHA personnel; and to personnel of other Federal, state, or local agencies with regulatory authority over the site.

(b)(2) Organizational structure part of the site program. (i) The organizational structure part of the program shall establish the specific chain of command and specify the overall responsibilities of supervisors and employees. It shall include, at a minimum, the following elements:

(b)(2)(i)(a) A general supervisor who has the responsibility and authority to direct all hazardous waste operations.

(b)(2)(i)(b) A site safety and health supervisor who has the responsibility and authority to develop and implement the site safety and health plan and verify compliance.

(b)(2)(i)(c) All other personnel needed for hazardous waste site operations and emergency response and their general functions and responsibilities.

(b)(2)(i)(d) The lines of authority, responsibility, and communication.

(b)(2)(ii) The organizational structure shall be reviewed and updated as necessary to reflect the current status of waste site operations.

(b)(3) Comprehensive workplan part of the site program. The comprehensive work plan part of the program shall address the tasks and objectives of the site operations and the logistics and resources required to reach those tasks and objectives.

(b)(3)(i) The comprehensive workplan shall address anticipated clean-up activities as well as normal operating procedures which need not repeat the employer's procedures available elsewhere.

(b)(3)(ii) The comprehensive workplan shall define work tasks and objectives and identify the methods for accomplishing those tasks and objectives.

(b)(3)(iii) The comprehensive workplan shall establish personnel requirements for implementing the plan.

(b)(3)(iv) The comprehensive workplan shall provide for the implementation of the training required in paragraph (e) of this section.

(b)(3)(v) The comprehensive workplan shall provide for the implementation of the informational programs required in paragraph (i) of this section.

(b)(3)(vi) The comprehensive workplan shall provide for the implementation of the medical surveillance program described in paragraph (f) of this section.

(b)(4) Site-specific safety and health plan part of the program. (i) General. The site safety and health plan, which must be kept on site, shall address the safety and health hazards of each phase of site operation and include requirements and procedures for employee protection.

(b)(4)(ii) Elements. The site safety and health plan, as a minimum, shall address the following:

(b)(4)(ii)(a) A safety and health risk or hazard analysis for each site task and operation found in the workplan.

(b)(4)(ii)(b) Employee training assignments to assure compliance with paragraph (e) of this section.

(b)(4)(ii)(c) Personal protective equipment to be used by employees for each of the site tasks and operations being conducted as required by the personal protective equipment program in paragraph (g)(5) of this section.

(b)(4)(ii)(d) Medical surveillance requirements in accordance with the program in paragraph (f) of this section.

(b)(4)(ii)(e) Frequency and types of air monitoring, personnel monitoring, and environmental sampling techniques and instrumentation to be used, including methods of maintenance and calibration of monitoring and sampling equipment to be used.

(b)(4)(ii)(f) Site control measures in accordance with the site control program required in paragraph (d) of this section.

(b)(4)(ii)(g) Decontamination procedures in accordance with paragraph (k) of this section.

(b)(4)(ii)(h) An emergency response plan meeting the requirements of paragraph (l) of this section for safe and effective responses to emergencies, including the necessary PPE and other equipment.

(b)(4)(ii)(i) Confined space entry procedures.

(b)(4)(ii)(j) A spill containment program meeting the requirements of paragraph (j) of this section.

(b)(4)(iii) Pre-entry briefing. The site specific safety and health plan shall provide for pre-entry briefings to be held prior to initiating any site activity, and at such other times as necessary to ensure that employees are apprised of the site safety and health plan and that this plan is being followed. The information and data obtained from site characterization and analysis work required in paragraph (c) of this section shall be used to prepare and update the site safety and health plan.

(b)(4)(iv) Effectiveness of site safety and health plan. Inspections shall be conducted by the site safety and health supervisor or, in the absence of that individual, another individual who is knowledgeable in occupational safety and health, acting on behalf of the employer as necessary to determine the effectiveness of the site safety and health plan. Any deficiencies in the effectiveness of the site safety and health plan shall be corrected by the employer.

(c) Site characterization and analysis— (1) General. Hazardous waste sites shall be evaluated in accordance with this paragraph to identify specific site hazards and to determine the appropriate safety and health control procedures needed to protect employees from the identified hazards.

(c)(2) Preliminary evaluation. A preliminary evaluation of a site's characteristics shall be performed prior to site entry by a qualified person in order to aid in the selection of appropriate employee protection methods prior to site entry. Immediately after initial site entry, a more detailed evaluation of the site's specific characteristics shall be performed by a qualified person in order to further identify existing site hazards and to further aid in the selection of the appropriate engineering controls and personal protective equipment for the tasks to be performed.

(c)(3) Hazard identification. All suspected conditions that may pose inhalation or skin absorption hazards that are immediately dangerous to life or health (IDLH), or other conditions that may cause death or serious harm, shall be identified during the preliminary survey and evaluated during the detailed survey. Examples of such hazards include, but are not limited to, confined space entry, potentially explosive or flammable situations, visible vapor clouds, or areas where biological indicators such as dead animals or vegetation are located.

(c)(4) Required information. The following information to the extent available shall be obtained by the employer prior to allowing employees to enter a site:

(c)(4)(i) Location and approximate size of the site.

(c)(4)(ii) Description of the response activity and/or the job task to be performed.

(c)(4)(iii) Duration of the planned employee activity.

(c)(4)(iv) Site topography and accessibility by air and roads.

(c)(4)(v) Safety and health hazards expected at the site.

(c)(4)(vi) Pathways for hazardous substance dispersion.

(c)(4)(vii) Present status and capabilities of emergency response teams that would provide assistance to hazardous waste clean-up site employees at the time of an emergency.

(c)(4)(viii) Hazardous substances and health hazards involved or expected at the site, and their chemical and physical properties.

(c)(5) Personal protective equipment. Personal protective equipment (PPE) shall be provided and used during initial site entry in accordance with the following requirements:

(c)(5)(i) Based upon the results of the preliminary site evaluation, an ensemble of PPE shall be selected and used during the initial site entry which will provide protection to a level of exposure below permissible exposure limits and published exposure levels for known or suspected hazardous substances and health hazards, and which will provide protection against other known and suspected hazards identified during the preliminary site evaluation. If there is no permissible exposure limit or published exposure level, the employer may use other published studies and information as a guide to appropriate personal protective equipment.

(c)(5)(ii) If positive-pressure self-contained breathing apparatus is not used as part of the entry ensemble, and if respiratory protection is warranted by the potential hazards identified during the preliminary site evaluation, an escape self-contained breathing apparatus of at least five minutes duration shall be carried by employees during initial site entry.

(c)(5)(iii) If preliminary site evaluation does not produce sufficient information to identify the hazard or suspected hazards of the site, an ensemble providing protection equivalent to Level B PPE shall be provided as minimum protection, and direct reading instruments shall be used as appropriate for identifying IDLH conditions. (See Appendix B for a description of Level B hazards and the recommendations for Level B protective equipment.)

(c)(5)(iv) Once the hazards of the site have been identified, the appropriate PPE shall be selected and used in accordance with paragraph (g) of this section.

(c)(6) Monitoring. The following monitoring shall be conducted during initial site entry when the site evaluation produces information that shows the potential for ionizing radiation or IDLH conditions, or when the site information is not sufficient reasonably to eliminate these possible conditions:

(c)(6)(i) Monitoring with direct reading instruments for hazardous levels of ionizing radiation.

(c)(6)(ii) Monitoring the air with appropriate direct reading test equipment (i.e., combustible gas meters, detector tubes) for IDLH and other conditions that may cause death or serious harm (combustible or explosive atmospheres, oxygen deficiency, toxic substances).

(c)(6)(iii) Visually observing for signs of actual or potential IDLH or other dangerous conditions.

(c)(6)(iv) An ongoing air monitoring program in accordance with paragraph (h) of this section shall be implemented after site characterization has determined the site is safe for the startup of operations.

(c)(7) Risk identification. Once the presence and concentrations of specific hazardous substances and health hazards have been established, the risks associated with these substance shall be identified. Employees who will be working on the site shall be informed of any risks that have been identified. In situations covered by the Hazard Communication Standard, 29 CFR 1910.1200, training required by that standard need not be duplicated.

Note to (c)(7).—Risks to consider include, but are not limited to:

(a) Exposure exceeding the permissible exposure limits and published exposure levels.

(b) IDLH concentrations.

(c) Potential skin absorption and irritation sources.

(d) Potential eye irritation sources.

(e) Explosion sensitivity and flammability ranges.

(f) Oxygen deficiency.

(c)(8) Employee notification. Any information concerning the physical, chemical, and toxicologic properties of each substance known or expected to be present on site that is available to the employer and relevant to the duties an employee is expected to perform shall be made available to the affected employees prior to the commencement of their work activities. The employer may utilize information developed for the hazard communication standard for this purpose.

(d) Site control—(1) General. Appropriate site control procedures shall be implemented to control employee exposure to hazardous substances before clean-up work begins.

(d)(2) Site control program. A site control program for protecting employees which is part of the employer's site safety and health program required in paragraph (b) of this section shall be developed during the planning stages of a hazardous waste clean-up operation and modified as necessary as new information becomes available.

(d)(3) Elements of the site control program. The site control program shall, as a minimum, include: A site map; site work zones; the use of a "buddy system"; site communications including alerting means for emergencies; the standard operating procedures or safe work practices; and identification of the nearest medical assistance. Where these requirements are covered elsewhere, they need not be repeated.

(e) Training—(1) General. (i) All employees working on site (such as but not limited to equipment operators, general laborers and others) exposed to hazardous substances, health hazards, or safety hazards and their supervisors and management responsible for the site shall receive training meeting the requirements of this paragraph before they are permitted to engage in hazardous waste operations that could expose them to hazardous substances, safety, or health hazards, and they shall receive review training as specified in this paragraph.

(e)(1)(ii) Employees shall not be permitted to participate in or supervise field activities until they have been trained to a level required by their job function and responsibility.

(e)(2) Elements to be covered. The training shall thoroughly cover the following:

(e)(2)(i) Names of personnel and alternates responsible for site safety and health;

(e)(2)(ii) Safety, health and other hazards present on the site;

(e)(2)(iii) Use of personal protective equipment;

(e)(2)(iv) Work practices by which the employee can minimize risks from hazards;

(e)(2)(v) Safe use of engineering controls and equipment on the site;

(e)(2)(vi) Medical surveillance requirements, including recognition of symptoms and signs which might indicate overexposure to hazards; and

(e)(2)(vii) The contents of paragraphs (g) through (j) of the site safety and health plan set forth in paragraph (b)(4)(ii) of this section.

(e)(3) Initial training. (i) General site workers (such as equipment operators, general laborers and supervisory personnel) engaged in hazardous substance removal or other activities which expose or potentially expose workers to hazardous substances and health hazards shall receive a minimum of 40 hours of instruction off the site, and a minimum of three days actual field experience under the direct supervision of a trained, experienced supervisor.

(e)(3)(ii) Workers on site only occasionally for a specific limited task (such as, but not limited to, ground water monitoring, land surveying, or geophysical surveying) and who are unlikely to be exposed over permissible exposure limits and published exposure limits shall receive a minimum of 24 hours of instruction off the site, and the minimum of one day actual field experience under the direct supervision of a trained, experienced supervisor.

(e)(3)(iii) Workers regularly on site who work in areas which have been monitored and fully characterized indicating that exposures are under permissible exposure limits and published exposure limits where respirators are not necessary, and the characterization indicates that there are no health hazards or the possibility of an emergency developing, shall receive a minimum of 24 hours of instruction off the site and the minimum of one day actual field experience under direct supervision of a trained, experienced supervisor.

(e)(3)(iv) Workers with 24 hours of training who are covered by paragraphs (e)(3)(ii) and (e)(3)(iii) of this section, and who become general site workers or who are required to wear respirators, shall have the additional 16 hours and two days of training necessary to total the training specified in paragraph (e)(3)(i).

[55 F.R. 14073, April 13, 1990]

(e)(4) Management and supervisor training. On-site management and supervisors directly responsible for, or who supervise employees engaged in, hazardous waste operations shall receive 40 hours of initial training, and three days of supervised field experience (the training may be reduced to 24 hours and one day if the only area of their responsibility is employees covered by paragraphs (e)(3)(ii) and (e)(3)(iii)) and at least eight additional hours of specialized training at the time of job assignment on such topics as, but not limited to, the employer's safety and health program and the associated employee training program, personal protective equipment program, spill containment program, and health hazard monitoring procedure and techniques.

(e)(5) Qualifications for trainers. Trainers shall be qualified to instruct employees about the subject matter that is being presented in training. Such trainers shall have satisfactorily completed a training program for teaching the subjects they are expected to teach, or they shall have the academic credentials and instructional experience necessary for teaching the subjects. Instructors shall demonstrate competent instructional skills and knowledge of the applicable subject matter.

(e)(6) Training certification. Employees and supervisors that have received and successfully completed the training and field experience specified in paragraphs (e)(1) through (e)(4) of this section shall be certified by their instructor or the head instructor and trained supervisor as having successfully completed the necessary training. A written certificate shall be given to each person so certified. Any person who has not been so certified or who does not meet the requirements of paragraph (e)(9) of this section shall be prohibited from engaging in hazardous waste operations.

(e)(7) Emergency response. Employees who are engaged in responding to hazardous emergency situations at hazardous waste clean-up sites that may expose them to hazardous substances shall be trained in how to respond to such expected emergencies.

(e)(8) Refresher training. Employees specified in paragraph (e)(1) of this section, and managers and supervisors specified in paragraph (e)(4) of this section, shall receive eight hours of refresher training annually on the items specified in paragraph (e)(2) and/or (e)(4) of this section, any critique of incidents that have occurred in the past year that can serve as training examples of related work, and other relevant topics.

(e)(9) Equivalent training. Employers who can show by documentation or certification that an employee's work experience and/or training has resulted in training equivalent to that training required in paragraphs (e)(1) through (e)(4) of this section shall not be required to provide the initial training requirements of those paragraphs to such employees. However, certified employees or employees with equivalent training new to a site shall receive appropriate, site specific training before site entry and have appropriate supervised field experience at the new site. Equivalent training includes any academic training or the training that existing employees might have already received from actual hazardous waste site work experience.

[55 F.R. 14073, April 13, 1990]

(f) Medical surveillance—(1) General. Employers engaged in operations specified in paragraphs (a)(1)(i) through (a)(1)(iv) of this section and not covered by (a)(2)(iii) exceptions and employers of employees specified in paragraph (q)(9) shall institute a medical surveillance program in accordance with this paragraph.

(f)(2) Employees covered. The medical surveillance program shall be instituted by the employer for the following employees:

(f)(2)(i) All employees who are or may be

exposed to hazardous substances or health hazards at or above the permissible exposure limits or, if there is no permissible exposure limit, above the published exposure levels for these substances, without regard to the use of respirators, for 30 days or more a year;

(f)(2)(ii) All employees who wear a respirator for 30 days or more a year or as required by §1910.134;

(f)(2)(iii) All employees who are injured, become ill or develop signs or symptoms due to possible overexposure involving hazardous substances or health hazards; from an emergency response or hazardous waste operation; and

[55 F.R. 14073, April 13, 1990]

(f)(2)(iv) Members of HAZMAT teams.

(f)(3) Frequency of medical examinations and consultations. Medical examinations and consultations shall be made available by the employer to each employee covered under paragraph (f)(2) of this section on the following schedules:

(f)(3)(i) For employees covered under paragraphs (f)(2)(i), (f)(2)(ii), and (f)(2)(iv):

(f)(3)(i)(a) Prior to assignment;

(f)(3)(i)(b) At least once every twelve months for each employee covered unless the attending physician believes a longer interval (not greater than biennially) is appropriate;

(f)(3)(i)(c) At termination of employment or reassignment to an area where the employee would not be covered if the employee has not had an examination within the last six months;

(f)(3)(i)(d) As soon as possible upon notification by an employee that the employee has developed signs or symptoms indicating possible overexposure to hazardous substances or health hazards, or that the employee has been injured or exposed above the permissible exposure limits or published exposure levels in an emergency situation;

(f)(3)(i)(e) At more frequent times, if the examining physician determines that an increased frequency of examination is medically necessary.

(f)(3)(ii) For employees covered under paragraph (f)(2)(iii) and for all employees including those of employers covered by paragraph (a)(1)(v) who may have been injured, received a health impairment, developed signs or symptoms which may have resulted from exposure to hazardous substances resulting from an emergency incident, or exposed during an emergency incident to hazardous substances at concentrations above the permissible exposure limits or the published exposure levels without the necessary personal protective equipment being used:

(f)(3)(ii)(a) As soon as possible following the emergency incident or development of signs or symptoms;

(f)(3)(ii)(b) At additional times, if the examining physician determines that follow-up examinations or consultations are medically necessary.

(f)(4) Content of medical examinations and consultations. (i) Medical examinations required by paragraph (f)(3) of this section shall include a medical and work history (or updated history if one is in the employee's file) with special emphasis on symptoms related to the handling of hazardous substances and health hazards, and to fitness for duty including the ability to wear any required PPE under conditions (i.e., temperature extremes) that may be expected at the work site.

(f)(4)(ii) The content of medical examinations or consultations made available to employees pursuant to paragraph (f) shall be determined by the attending physician. The guidelines in the *Occupational Safety and Health Guidance Manual for Hazardous Waste Site Activities* (see Appendix D, Reference #10) should be consulted.

(f)(5) Examinations by a physician and costs. All medical examinations and procedures shall be performed by or under the supervision of a licensed physician, preferably one knowledgeable in occupational medicine, and shall be provided without cost to the employee, without loss of pay, and at a reasonable time and place.

(f)(6) Information provided to the physician. The employer shall provide one copy of this standard and its appendices to the attending physician, and in addition, the following for each employee:

(f)(6)(i) A description of the employee's duties as they relate to the employee's exposures.

(f)(6)(ii) The employee's exposure levels or anticipated exposure levels.

(f)(6)(iii) A description of any personal protective equipment used or to be used.

(f)(6)(iv) Information from previous medical examinations of the employee which is not readily available to the examining physician.

(f)(6)(v) Information required by §1910.134.

(f)(7) Physician's written opinion. (i) The employer shall obtain and furnish the employee with a copy of a written opinion from the attending physician containing the following:

(f)(7)(i)(a) The physician's opinion as to whether the employee has any detected medical conditions which would place the employee at increased risk of material impairment of the employee's health from work in hazardous waste operations or emergency response, or from respirator use.

(f)(7)(i)(b) The physician's recommended limitations upon the employee's assigned work.

(f)(7)(i)(c) The results of the medical examination and tests if requested by the employee.

(f)(7)(i)(d) A statement that the employee has been informed by the physician of the results of the medical examination and any medical conditions which require further examination or treatment.

(f)(7)(ii) The written opinion obtained by the employer shall not reveal specific findings or diagnoses unrelated to occupational exposures.

(f)(8) Recordkeeping. (i) An accurate record of the medical surveillance required by paragraph (f) of this section shall be retained. This record shall be retained for the period specified and meet the criteria of 29 CFR 1910.20.

(f)(8)(ii) The record required in paragraph (f)(8)(i) of this section shall include at least the following information:

(f)(8)(ii)(a) The name and social security number of the employee;

(f)(8)(ii)(b) Physician's written opinions, recommended limitations, and results of examinations and tests;

(f)(8)(ii)(c) Any employee medical complaints related to exposure to hazardous substances;

(f)(8)(ii)(d) A copy of the information provided to the examining physician by the employer, with the exception of the standard and its appendices.

(g) Engineering controls, work practices, and personal protective equipment for employee protection—Engineering controls, work practices, personal protective equipment, or a combination of these shall be implemented in accordance with this paragraph to protect employees from exposure to hazardous substances and safety and health hazards.

(g)(1) Engineering controls, work practices and PPE for substances regulated in Subparts G and Z. (i) Engineering controls and work practices shall be instituted to reduce and maintain employee exposure to or below the permissible exposure limits for substances regulated by 29 CFR Part 1910, to the extent required by Subpart Z, except to the extent that such controls and practices are not feasible.

Note to (g)(1)(i): Engineering controls which may be feasible include the use of pressurized cabs or control booths on equipment, and/or the use of remotely operated material handling equipment. Work practices which may be feasible are removing all non-essential employees from potential exposure during opening of drums, wetting down dusty operations and locating employees upwind of possible hazards.

(g)(1)(ii) Whenever engineering controls and work practices are not feasible or not required, any reasonable combination of engineering controls, work practices and PPE shall be used.

[55 F.R. 14073, April 13, 1990]

(g)(1)(iii) The employer shall not implement a schedule of employee rotation as a means of compliance with permissible exposure limits or dose limits except when there is no other feasible way of complying with the airborne or dermal dose limits for ionizing radiation.

(g)(1)(iv) The provisions of 29 CFR, Subpart G, shall be followed.

(g)(2) Engineering controls, work practices and PPE for substances not regulated in Subparts G and Z. An appropriate combination of engineering controls, work practices and personal protective equipment shall be used to reduce and maintain employee exposure to or below published exposure levels for hazardous substances and health hazards not regulated by 29 CFR Part 1910, Subparts G and Z. The employer may use the published literature and MSDS as a guide in making the employer's determination as to what level of protection the employer believes is appropriate for hazardous substances and health hazards for which there is no permissible exposure limit or published exposure limit.

(g)(3) Personal protective equipment selection. (i) Personal protective equipment (PPE) shall be selected and used which will protect employees from the hazards and potential hazards they are likely to encounter as identified during the site characterization and analysis.

(g)(3)(ii) Personal protective equipment selection shall be based on an evaluation of the performance characteristics of the PPE relative to the requirements and limitations of the site, the task-specific conditions and duration, and the hazards and potential hazards identified at the site.

(g)(3)(iii) Positive pressure self-contained breathing apparatus, or positive pressure airline respirators equipped with an escape air supply, shall be used when chemical exposure levels present will create a substantial possibility of immediate death, immediate serious illness or injury, or impair the ability to escape.

(g)(3)(iv) Totally-encapsulating chemical protective suits (protection equivalent to Level A protection as recommended in Appendix B) shall be used in conditions where skin absorption of a hazardous substance may result in a substantial possibility of immediate death, immediate serious illness or injury, or impair the ability to escape.

(g)(3)(v) The level of protection provided by PPE selection shall be increased when additional information on site conditions indicates that increased protection is necessary to reduce employee exposures below permissible exposure limits and published exposure levels for hazardous substances and health hazards. (See Appendix B for guidance on selecting PPE ensembles.)

Note to (g)(3): The level of employee protection provided may be decreased when additional information or site conditions show that decreased protection will not result in hazardous exposures to employees.

(g)(3)(vi) Personal protective equipment shall be selected and used to meet the requirements of 29 CFR Part 1910, Subpart I, and additional requirements specified in this section.

(g)(4) Totally-encapsulating chemical protective suits. (i) Totally-encapsulating suits shall protect employees from the particular hazards which are identified during site characterization and analysis.

(g)(4)(ii) Totally-encapsulating suits shall be capable of maintaining positive air pressure. (See Appendix A for a test method which may be used to evaluate this requirement.)

(g)(4)(iii) Totally-encapsulating suits shall be capable of preventing inward test gas leakage of more than 0.5 percent. (See Appendix A for a test method which may be used to evaluate this requirement.)

(g)(5) Personal protective equipment (PPE) program. A written personal protective equipment program, which is part of the employer's safety and health program required in paragraph (b) of this section or required in paragraph (p)(1) of this section and which is also a part of the site-specific safety and health plan shall be established. The PPE program shall address the elements listed below. When elements, such as donning and doffing procedures, are provided by the manufacturer of a piece of equipment and are attached to the plan, they need not be rewritten into the plan as long as they adequately address the procedure or element.

(g)(5)(i) PPE selection based upon site hazards,

(g)(5)(ii) PPE use and limitations of the equipment,

(g)(5)(iii) Work mission duration,

(g)(5)(iv) PPE maintenance and storage,

(g)(5)(v) PPE decontamination and disposal,

(g)(5)(vi) PPE training and proper fitting,

(g)(5)(vii) PPE donning and doffing procedures,

(g)(5)(viii) PPE inspection procedures prior to, during, and after use,

(g)(5)(ix) Evaluation of the effectiveness of the PPE program, and

(g)(5)(x) Limitations during temperature extremes, heat stress, and other appropriate medical considerations.

(h) Monitoring—(1) General. (i) Monitoring shall be performed in accordance with this paragraph where there may be a question of employee exposure to hazardous concentrations of hazardous substances in order to assure proper selection of engineering controls, work practices and personal protective equipment so that employees are not exposed to levels which exceed permissible exposure limits or published exposure levels if there are no permissible exposure limits, for hazardous substances.

[55 F.R. 14073, April 13, 1990]

(h)(1)(ii) Air monitoring shall be used to identify and quantify airborne levels of hazardous substances and safety and health hazards in order to determine the appropriate level of employee protection needed on site.

(h)(2) Initial entry. Upon initial entry, representative air monitoring shall be conducted to identify an IDLH condition, exposure over permissible exposure limits or published exposure levels, exposure over a radioactive material's dose limits or other dangerous condition such as the presence of flammable atmospheres or oxygen-deficient environments.

(h)(3) Periodic monitoring. Periodic monitoring shall be conducted when the possibility of an IDLH condition or flammable atmosphere has developed or when there is indication that exposures may have risen over permissible exposure limits or published exposure levels since prior monitoring. Situations where it shall be considered whether the possibility that exposures have risen are as follows:

(h)(3)(i) When work begins on a different portion of the site.

(h)(3)(ii) When contaminants other than those previously identified are being handled.

(h)(3)(iii) When a different type of operation is initiated (e.g., drum opening as opposed to exploratory well drilling).

(h)(3)(iv) When employees are handling leaking drums or containers or working in areas with obvious liquid contamination (e.g., a spill or lagoon).

(h)(4) Monitoring of high-risk employees. After the actual clean-up phase of any hazardous waste operation commences; for example, when soil, surface water or containers are moved or disturbed; the employer shall monitor those employees likely to have the highest exposure to hazardous substances and health hazards likely to be present above permissible exposure limits or published exposure levels by using personal sampling frequently enough to characterize employee exposures. If the employees likely to have the highest exposure are over permissible exposure limits or published exposure limits, then monitoring shall continue to determine all employees likely to be above those limits. The employer may utilize a representative sampling approach by documenting that the employees and chemicals chosen for monitoring are based on the criteria stated above.

Note to (h): It is not required to monitor employees engaged in site characterization operations covered by paragraph (c) of this section.

(i) Informational programs—Employers shall develop and implement a program, which is part of the employer's safety and health program required in paragraph (b) of this section, to inform employees, contractors, and subcontractors (or their representative) actually engaged in hazardous waste operations of the nature, level and degree of exposure likely as a result of participation in such hazardous waste operations. Employees, contractors and subcontractors working outside of the operations part of a site are not covered by this standard.

(j) Handling drums and containers— (1) General. (i) Hazardous substances and contaminated soils, liquids, and other residues shall be handled, transported, labeled, and disposed of in accordance with this paragraph.

(j)(1)(ii) Drums and containers used during the clean-up shall meet the appropriate DOT, OSHA, and EPA regulations for the wastes they contain.

(j)(1)(iii) When practical, drums and containers shall be inspected and their integrity shall be assured prior to being moved. Drums or containers that cannot be inspected before being moved because of storage conditions (i.e., buried beneath the earth, stacked behind other drums, stacked several tiers high in a pile, etc.) shall be moved to an accessible location and inspected prior to further handling.

(j)(1)(iv) Unlabeled drums and containers shall be considered to contain hazardous substances and handled accordingly until the contents are positively identified and labeled.

(j)(1)(v) Site operations shall be organized to minimize the amount of drum or container movement.

(j)(1)(vi) Prior to movement of drums or containers, all employees exposed to the transfer operation shall be warned of the potential hazards associated with the contents of the drums or containers.

(j)(1)(vii) U.S. Department of Transportation specified salvage drums or containers and suitable quantities of proper absorbent shall be kept available and used in areas where spills, leaks, or ruptures may occur.

(j)(1)(viii) Where major spills may occur, a spill containment program, which is part of the employer's safety and health program required in paragraph (b) of this section, shall be implemented to contain and isolate the entire volume of the hazardous substance being transferred.

(j)(1)(ix) Drums and containers that cannot be moved without rupture, leakage, or spillage shall be emptied into a sound container using a device classified for the material being transferred.

(j)(1)(x) A ground-penetrating system or other type of detection system or device shall be used to estimate the location and depth of buried drums or containers.

(j)(1)(xi) Soil or covering material shall be removed with caution to prevent drum or container rupture.

(j)(1)(xii) Fire extinguishing equipment meeting the requirements of 29 CFR Part 1910, Subpart L, shall be on hand and ready for use to control incipient fires.

(j)(2) Opening drums and containers. The following procedures shall be followed in areas where drums or containers are being opened:

(j)(2)(i) Where an airline respirator system is used, connections to the source of air supply shall be protected from contamination and the entire system shall be protected from physical damage.

(j)(2)(ii) Employees not actually involved in opening drums or containers shall be kept a safe distance from the drums or containers being opened.

(j)(2)(iii) If employees must work near or adjacent to drums or containers being opened, a suitable shield that does not interfere with the work operation shall be placed between the employee and the drums or containers being opened to protect the employee in case of accidental explosion.

(j)(2)(iv) Controls for drum or container opening equipment, monitoring equipment, and fire suppression equipment shall be located behind the explosion-resistant barrier.

(j)(2)(v) When there is a reasonable possibility of flammable atmospheres being present, material handling equipment and hand tools shall be of the type to prevent sources of ignition.

(j)(2)(vi) Drums and containers shall be opened in such a manner that excess interior pressure will be safely relieved. If pressure cannot be relieved from a remote location, appropriate shielding shall be placed between the employee and the drums or containers to reduce the risk of employee injury.

(j)(2)(vii) Employees shall not stand upon or work from drums or containers.

(j)(3) Material handling equipment. Material handling equipment used to transfer drums and containers shall be selected, positioned and operated to minimize the sources of ignition related to the equipment from igniting vapors released from ruptured drums or containers.

(j)(4) Radioactive wastes. Drums and containers containing radioactive wastes shall not be handled until such time as their hazard to employees is properly assessed.

(j)(5) Shock sensitive wastes. As a minimum, the following special precautions shall be taken when drums and containers containing or suspected of containing shock-sensitive wastes are handled:

(j)(5)(i) All non-essential employees shall be evacuated from the area of transfer.

(j)(5)(ii) Material handling equipment shall be provided with explosive containment devices or protective shields to protect equipment operators from exploding containers.

(j)(5)(iii) An employee alarm system capable of being perceived above surrounding light and noise conditions shall be used to signal the commencement and completion of explosive waste handling activities.

(j)(5)(iv) Continuous communications (i.e., portable radios, hand signals, telephones, as appropriate) shall be maintained between the employee-in-charge of the immediate handling area and both the site safety and health supervisor and the command post until such time as the handling operation is completed. Communication equipment or methods that could cause shock sensitive materials to explode shall not be used.

(j)(5)(v) Drums and containers under pressure, as evidenced by bulging or swelling, shall not be moved until such time as the cause for excess pressure is determined and appropriate containment procedures have been implemented to protect employees from explosive relief of the drum.

(j)(5)(vi) Drums and containers containing packaged laboratory wastes shall be considered to contain shock-sensitive or explosive materials until they have been characterized.

Caution: Shipping of shock sensitive wastes may be prohibited under U.S. Department of Transportation regulations. Employers and their shippers should refer to 49 CFR 173.21 and 173.50.

(j)(6) Laboratory waste packs. In addition to the requirements of paragraph (j)(5) of this section, the following precautions shall be taken, as a minimum, in handling laboratory waste packs (lab packs):

(j)(6)(i) Lab packs shall be opened only when necessary and then only by an individual knowledgeable in the inspection, classification, and segregation of the containers within the pack according to the hazards of the wastes.

(j)(6)(ii) If crystalline material is noted on any container, the contents shall be handled as a shock-sensitive waste until the contents are identified.

(j)(7) Sampling of drum and container contents. Sampling of containers and drums shall be done in accordance with a sampling procedure which is part of the site safety and health plan developed for and available to employees and others at the specific worksite.

(j)(8) Shipping and transport. (i) Drums and containers shall be identified and classified prior to packaging for shipment.

(j)(8)(ii) Drum or container staging areas shall be kept to the minimum number necessary to identify and classify materials safely and prepare them for transport.

(j)(8)(iii) Staging areas shall be provided with adequate access and egress routes.

(j)(8)(iv) Bulking of hazardous wastes shall be permitted only after a thorough characterization of the materials has been completed.

(j)(9) Tank and vault procedures. (i) Tanks and vaults containing hazardous substances shall be handled in a manner similar to that for drums and containers, taking into consideration the size of the tank or vault.

(j)(9)(ii) Appropriate tank or vault entry procedures as described in the employer's safety and health plan shall be followed whenever employees must enter a tank or vault.

(k) Decontamination—(1) General. Procedures for all phases of decontamination shall be developed and implemented in accordance with this paragraph.

(k)(2) Decontamination procedures. (i) A decontamination procedure shall be developed, communicated to employees and implemented before any employees or equipment may enter areas on site where potential for exposure to hazardous substances exists.

(k)(2)(ii) Standard operating procedures shall be developed to minimize employee contact with hazardous substances or with equipment that has contacted hazardous substances.

(k)(2)(iii) All employees leaving a contaminated area shall be appropriately decontaminated; all contaminated clothing and equipment leaving a contaminated area shall be appropriately disposed of or decontaminated.

(k)(2)(iv) Decontamination procedures shall be monitored by the site safety and health supervisor to determine their effectiveness. When such procedures are found to be ineffective, appropriate steps shall be taken to correct any deficiencies.

(k)(3) Location. Decontamination shall be performed in geographical areas that will minimize the exposure of uncontaminated employees or equipment to contaminated employees or equipment.

(k)(4) Equipment and solvents. All equipment

and solvents used for decontamination shall be decontaminated or disposed of properly.

(k)(5) Personal protective clothing and equipment. (i) Protective clothing and equipment shall be decontaminated, cleaned, laundered, maintained or replaced as needed to maintain their effectiveness.

(k)(5)(ii) Employees whose non-impermeable clothing becomes wetted with hazardous substances shall immediately remove that clothing and proceed to shower. The clothing shall be disposed of or decontaminated before it is removed from the work zone.

(k)(6) Unauthorized employees. Unauthorized employees shall not remove protective clothing or equipment from change rooms.

(k)(7) Commercial laundries or cleaning establishments. Commercial laundries or cleaning establishments that decontaminate protective clothing or equipment shall be informed of the potentially harmful effects of exposures to hazardous substances.

(k)(8) Showers and change rooms. Where the decontamination procedure indicates a need for regular showers and change rooms outside of a contaminated area, they shall be provided and meet the requirements of 29 CFR 1910.141. If temperature conditions prevent the effective use of water, then other effective means for cleansing shall be provided and used.

(l) Emergency response by employees at uncontrolled hazardous waste sites—(1) Emergency response plan. (i) An emergency response plan shall be developed and implemented by all employers within the scope of this paragraph and paragraphs (a)(1)(i)–(ii) of this section to handle anticipated emergencies prior to the commencement of hazardous waste operations. The plan shall be in writing and available for inspection and copying by employees, their representatives, OSHA personnel and other governmental agencies with relevant responsibilities.

[55 F.R. 14073, April 13, 1990]

(l)(1)(ii) Employers who will evacuate their employees from the danger area when an emergency occurs, and who do not permit any of their employees to assist in handling the emergency, are exempt from the requirements of this paragraph if they provide an emergency action plan complying with section 1910.38(a) of this part.

[55 F.R. 14073, April 13, 1990]

(l)(2) Elements of an emergency response plan. The employer shall develop an emergency response plan for emergencies which shall address, as a minimum, the following:

(l)(2)(i) Pre-emergency planning.

(l)(2)(ii) Personnel roles, lines of authority, and communication.

(l)(2)(iii) Emergency recognition and prevention.

(l)(2)(iv) Safe distances and places of refuge.

(l)(2)(v) Site security and control.

(l)(2)(vi) Evacuation routes and procedures.

(l)(2)(vii) Decontamination procedures which are not covered by the site safety and health plan.

(l)(2)(viii) Emergency medical treatment and first aid.

(l)(2)(ix) Emergency alerting and response procedures.

(l)(2)(x) Critique of response and follow-up.

(l)(2)(xi) PPE and emergency equipment.

(l)(3) Procedures for handling emergency incidents. (i) In addition to the elements for the emergency response plan required in paragraph (l)(2) of this section, the following elements shall be included for emergency response plans:

(l)(3)(i)(a) Site topography, layout, and prevailing weather conditions.

(l)(3)(i)(b) Procedures for reporting incidents to local, state, and federal governmental agencies.

(l)(3)(ii) The emergency response plan shall be a separate section of the site safety and health plan.

(l)(3)(iii) The emergency response plan shall be compatible and integrated with the disaster, fire and/or emergency response plans of local, state, and federal agencies.

(l)(3)(iv) The emergency response plan shall be rehearsed regularly as part of the overall training program for site operations.

(l)(3)(v) The site emergency response plan shall be reviewed periodically and, as necessary, be amended to keep it current with new or changing site conditions or information.

(l)(3)(vi) An employee alarm system shall be installed in accordance with 29 CFR 1910.165 to notify employees of an emergency situation; to stop work activities if necessary; to lower background noise in order to speed communication; and to begin emergency procedures.

(l)(3)(vii) Based upon the information available at time of the emergency, the employer shall evaluate the incident and the site response capabilities and proceed with appropriate steps to implement the site emergency response plan.

(m) Illumination—Areas accessible to employees shall be lighted to not less than the minimum illumination intensities listed in the following Table H-120.1 while any work is in progress:

TABLE H-120.1 Minimum Illumination Intensities in Foot-Candles

Foot-Candles	Area or Operations
5	General site areas.
3	Excavation and waste areas, access ways, active storage areas, loading platforms, refueling, and field maintenance areas.
5	Indoors: warehouses, corridors, hallways, and exitways.
5	Tunnels, shafts, and general underground work areas. (Exception: Minimum of 10 foot-candles is required at tunnel and shaft heading during drilling, mucking, and scaling. Mine Safety and Health Administration approved cap lights shall be acceptable for use in the tunnel heading.)
10	General shops (e.g., mechanical and electrical equipment rooms, active storerooms, barracks or living quarters, locker or dressing rooms, dining areas, indoor toilets and workrooms).
30	First aid stations, infirmaries, and (n)(1)(ii) Portable containers used to dispense drinking water shall be capable of being tightly closed, and equipped with a tap. Water shall not be dipped from containers.

(n) Sanitation at temporary workplaces—(1) Potable water. (i) An adequate supply of potable water shall be provided on the site.

(n)(1)(ii) Portable containers used to dispense drinking water shall be tightly closed, and equipped with a tap. Water shall not be dipped from containers.

(n)(1)(iii) Any container used to distribute drinking water shall be clearly marked as to the nature of its contents and not used for any other purpose.

(n)(1)(iv) Where single service cups (to be used but once) are supplied, both a sanitary container for the unused cups and a receptacle for disposing of the used cups shall be provided.

(n)(2) Nonpotable water. (i) Outlets for nonpotable water, such as water for firefighting purposes, shall be identified to indicate clearly that the water is unsafe and is not to be used for drinking, washing, or cooking purposes.

(n)(2)(ii) There shall be no cross-connection, open or potential, between a system furnishing potable water and a system furnishing nonpotable water.

(n)(3) Toilet facilities. (i) Toilets shall be provided for employees according to the following Table H-120.2:

TABLE H-120.2 Toilet Facilities

Number of Employees	Minimum Number of Facilities
20 or fewer	One
More than 20, but fewer than 200	One toilet seat and one urinal per 40 employees
More than 200	One toilet seat and one urinal per 50 employees

(n)(3)(ii) Under temporary field conditions, provisions shall be made to assure that at least one toilet facility is available.

(n)(3)(iii) Hazardous waste sites not provided with a sanitary sewer shall be provided with the following toilet facilities unless prohibited by local codes:

(n)(3)(iii)(a) Chemical toilets;

(n)(3)(iii)(b) Recirculating toilets;

(n)(3)(iii)(c) Combustion toilets; or

(n)(3)(iii)(d) Flush toilets.

(n)(3)(iv) The requirements of this paragraph for sanitation facilities shall not apply to mobile crews having transportation readily available to nearby toilet facilities.

(n)(3)(v) Doors entering toilet facilities shall be provided with entrance locks controlled from inside the facility.

(n)(4) Food handling. All food service facilities and operations for employees shall meet the applicable laws, ordinances, and regulations of the jurisdictions in which they are located.

(n)(5) Temporary sleeping quarters. When temporary sleeping quarters are provided, they shall be heated, ventilated, and lighted.

(n)(6) Washing facilities. The employer shall provide adequate washing facilities for employees engaged in operations where hazardous substances may be harmful to employees. Such facilities shall be in near proximity to the worksite, in areas where exposures are below permissible exposure limits and published exposure levels and which are under the controls of the employer; and shall be so equipped as to enable employees to remove hazardous substances from themselves.

(n)(7) Showers and change rooms. When hazardous waste clean-up or removal operations commence on a site and the duration of the work will require six months or greater time to complete, the employer shall provide showers and change rooms for all employees exposed to hazardous substances and health hazards involved in hazardous waste clean-up or removal operations.

(n)(7)(i) Showers shall be provided and

shall meet the requirements of 29 CFR 1910.141(d)(3).

(n)(7)(ii) Change rooms shall be provided and shall meet the requirements of 29 CFR 1910.141(e). Change rooms shall consist of two separate change areas separated by the shower area required in paragraph (n)(7)(i) of this section. One change area, with an exit leading off the worksite, shall provide employees with a clean area where they can remove, store, and put on street clothing. The second area, with an exit to the worksite, shall provide the employees with an area where they can put on, remove and store work clothing and personal protective equipment.

(n)(7)(iii) Showers and change rooms shall be located in areas where exposures are below the permissible exposure limits and published exposure levels. If this cannot be accomplished, then a ventilation system shall be provided that will supply air that is below the permissible exposure limits and published exposure levels.

(n)(7)(iv) Employers shall assure that employees shower at the end of their work shift and when leaving the hazardous waste site.

(o) New technology programs—(1) The employer shall develop and implement procedures for the introduction of effective new technologies and equipment developed for the improved protection of employees working with hazardous waste clean-up operations, and the same shall be implemented as part of the site safety and health program to assure that employee protection is being maintained.

(o)(2) New technologies, equipment or control measures available to the industry, such as the use of foams, absorbents, adsorbents, neutralizers, or other means to suppress the level of air contaminates while excavating the site or for spill control, shall be evaluated by employers or their representatives. Such an evaluation shall be done to determine the effectiveness of new methods, materials, or equipment before implementing their use on a large scale for enhancing employee protection. Information and data from manufacturers or suppliers may be used as part of the employer's evaluation effort. Such evaluations shall be made available to OSHA upon request.

(p) Certain operations conducted under the Resource Conservation and Recovery Act of 1976 (RCRA)—Employers conducting operations at treatment, storage and disposal (TSD) facilities specified in paragraph (a)(1)(iv) of this section shall provide and implement the program specified in this paragraph. See the "Notes and Exceptions" to paragraph (a)(2)(iii) of this section for employers not covered.

[55 F.R. 14073, April 13, 1990]

(p)(1) Safety and health program. The employer shall develop and implement a written safety and health program for employees involved in hazardous waste operations that shall be available for inspection by employees, their representatives and OSHA personnel. The program shall be designed to identify, evaluate and control safety and health hazards in their facilities for the purpose of employee protection, to provide for emergency response meeting the requirements of paragraph (p)(8) of this section and to address as appropriate site analysis, engineering controls, maximum exposure limits, hazardous waste handling procedures and uses of new technologies.

(p)(2) Hazard communication program. The employer shall implement a hazard communication program meeting the requirements of 29 CFR 1910.1200 as part of the employer's safety and health program.

Note to 1910.120. The exemption for hazardous waste provided in §1910.1200 is applicable to this section.

(p)(3) Medical surveillance program. The employer shall develop and implement a medical surveillance program meeting the requirements of paragraph (f) of this section.

(p)(4) Decontamination program. The employer shall develop and implement a decontamination procedure meeting the requirements of paragraph (k) of this section.

(p)(5) New technology program. The employer shall develop and implement procedures meeting the requirements of paragraph (o) of this section for introducing new and innovative equipment into the workplace.

(p)(6) Material handling program. Where employees will be handling drums or containers, the employer shall develop and implement procedures meeting the requirements of paragraphs (j)(1)(ii) through (viii) and (xi) of this section, as well as (j)(3) and (j)(8) of this section prior to starting such work.

(p)(7) Training program. (i) New employees. The employer shall develop and implement a training program, which is part of the employer's safety and health program, for employees exposed to health hazards or hazardous substances at TSD operations to enable the employees to perform their assigned duties and functions in a safe and healthful manner so as not to endanger themselves or other employees. The initial training shall be for 24 hours and refresher training shall be for eight hours annually. Employees who have received the initial training required by this paragraph shall be given a written certificate attesting that they have successfully completed the necessary training.

[55 F.R. 14074, April 13, 1990]

(p)(7)(ii) Current employees. Employers who can show by an employee's previous work experience and/or training that the employee has had training equivalent to the initial training required by this paragraph shall be considered as meeting the initial training requirements of this paragraph as to that employee. Equivalent training includes the training that existing employees might have already received from actual site work experience. Current employees shall receive eight hours of refresher training annually.

(p)(7)(iii) Trainers. Trainers who teach initial training shall have satisfactorily completed a training course for teaching the subjects they are expected to teach or they shall have the academic credentials and instruction experience necessary to demonstrate a good command of the subject matter of the courses and competent instructional skills.

(p)(8) Emergency response program. (i) Emergency response plan. An emergency response plan shall be developed and implemented by all employers. Such plans need not duplicate any of the subjects fully addressed in the employer's contingency planning required by permits, such as those issued by the U.S. Environmental Protection Agency, provided that the contingency plan is made part of the emergency response plan. The emergency response plan shall be a written portion of the employer's safety and health program required in paragraph (p)(1) of this section. Employers who will evacuate their employees from the worksite location when an emergency occurs and who do not permit any of their employees to assist in handling the emergency are exempt from the requirements of paragraph (p)(8) if they provide an emergency action plan complying with §1910.38(a) of this part.

(p)(8)(ii) Elements of an emergency response plan. The employer shall develop an emergency response plan for emergencies which shall address, as a minimum, the following areas to the extent that they are not addressed in any specific program required in this paragraph:

(p)(8)(ii)(a) Pre-emergency planning and coordination with outside parties.

(p)(8)(ii)(b) Personnel roles, lines of authority, and communication.

(p)(8)(ii)(c) Emergency recognition and prevention.

(p)(8)(ii)(d) Safe distances and places of refuge.

(p)(8)(ii)(e) Site security and control.

(p)(8)(ii)(f) Evacuation routes and procedures.

(p)(8)(ii)(g) Decontamination procedures.

(p)(8)(ii)(h) Emergency medical treatment and first aid.

(p)(8)(ii)(i) Emergency alerting and response procedures.

(p)(8)(ii)(j) Critique of response and follow-up.

(p)(8)(ii)(k) PPE and emergency equipment.

(p)(8)(iii) Training. (a) Training for emergency response employees shall be completed before they are called upon to perform in real emergencies. Such training shall include the elements of the emergency response plan, standard operating procedures the employer has established for the job, the personal protective equipment to be worn and procedures for handling emergency incidents.

Exception #1: An employer need not train all employees to the degree specified if the employer divides the work force in a manner such that a sufficient number of employees who have responsibility to control emergencies have the training specified, and all other employees, who may first respond to an emergency incident, have sufficient awareness training to recognize that an emergency response situation exists and that they are instructed in that case to summon the fully trained employees and not attempt to control activities for which they are not trained.

Exception #2: An employer need not train all employees to the degree specified if arrangements have been made in advance for an outside fully-trained emergency response team to respond in a reasonable period and all employees, who may come to the incident first, have sufficient awareness training to recognize that an emergency response situation exists and they have been instructed to call the designated outside fully-trained emergency response team for assistance.

(p)(8)(iii)(b) Employee members of TSD facility emergency response organizations shall be trained to a level of competence in the recognition of health and safety hazards to protect themselves and other employees. This would include training in the methods used to minimize the risk from safety and health hazards; in the safe use of control equipment; in the selection and use of appropriate personal protective equipment; in the safe operating procedures to be used at the incident scene; in the techniques of coordination with other employees to minimize risks; in the appropriate response to over exposure from health hazards or injury to themselves and other employees; and in the recognition of subsequent symptoms which may result from over exposures.

(p)(8)(iii)(c) The employer shall certify that each covered employee has attended and successfully completed the training required in paragraph (p)(8)(iii) of this section, or shall certify the employee's competency at least yearly. The method used to demonstrate competency for certification of training shall be recorded and maintained by the employer.

(p)(8)(iv) Procedures for handling emergency incidents. (a) In addition to the elements for the emergency response plan required in paragraph (p)(8)(ii) of this section, the following elements shall be included for emergency response plans to the extent that they do not repeat any information already contained in the emergency response plan:

(p)(8)(iv)(a)(1) Site topography, layout, and prevailing weather conditions.

(p)(8)(iv)(a)(2) Procedures for reporting incidents to local, state, and federal governmental agencies.

(p)(8)(iv)(b) The emergency response plan shall be compatible and integrated with the disaster, fire and/or emergency response plans of local, state, and federal agencies.

(p)(8)(iv)(c) The emergency response plan shall be rehearsed regularly as part of the overall training program for site operations.

(p)(8)(iv)(d) The site emergency response plan shall be reviewed periodically and, as necessary, be amended to keep it current with new or changing site conditions or information.

(p)(8)(iv)(e) An employee alarm system shall be installed in accordance with 29 CFR 1910.165 to notify employees of an emergency situation; to stop work activities if necessary; to lower background noise in order to speed communication; and to begin emergency procedures.

(p)(8)(iv)(f) Based upon the information available at time of the emergency, the employer shall evaluate the incident and the site response capabilities and proceed with the appropriate steps to implement the site emergency response plan.

(q) Emergency response to hazardous substance releases—This paragraph covers employers whose employees are engaged in emergency response no matter where it occurs except that it does not cover employees engaged in operations specified in paragraphs (a)(1)(i) through (a)(1)(iv) of this section. Those emergency response organizations who have developed and implemented programs equivalent to this paragraph for handling releases of hazardous substances pursuant to section 303 of the Superfund Amendments and Reauthorization Act of 1986 (Emergency Planning and Community Right-to-Know Act of 1986, 42 U.S.C. 11003) shall be deemed to have met the requirements of this paragraph.

(q)(1) Emergency response plan. An emergency response plan shall be developed and implemented to handle anticipated emergencies prior to the commencement of emergency response operations. The plan shall be in writing and available for inspection and copying by employees, their representatives and OSHA personnel. Employers who will evacuate their employees from the danger area when an emergency occurs, and who do not permit any of their employees to assist in handling the emergency, are exempt from the requirements of this paragraph if they provide an emergency action plan in accordance with §1910.38(a) of this part.

(q)(2) Elements of an emergency response plan. The employer shall develop an emergency response plan for emergencies which shall address, as a minimum, the following to the extent that they are not addressed elsewhere:

(q)(2)(i) Pre-emergency planning and coordination with outside parties.

(q)(2)(ii) Personnel roles, lines of authority, training, and communication.

(q)(2)(iii) Emergency recognition and prevention.

(q)(2)(iv) Safe distances and places of refuge.

(q)(2)(v) Site security and control.

(q)(2)(vi) Evacuation routes and procedures.

(q)(2)(vii) Decontamination.

(q)(2)(viii) Emergency medical treatment and first aid.

(q)(2)(ix) Emergency alerting and response procedures.

(q)(2)(x) Critique of response and follow-up.

(q)(2)(xi) PPE and emergency equipment.

(q)(2)(xii) Emergency response organizations may use the local emergency response plan or the state emergency response plan, or both, as part of their emergency response plan to avoid duplication. Those items of the emergency response plan that are being properly addressed by the SARA Title III plans may be substituted into their emergency plan or otherwise kept together for the employer and employee's use.

(q)(3) Procedures for handling emergency response. (i) The senior emergency official responding to an emergency shall become the individual in charge of a site-specific Incident Command System (ICS). All emergency responders and their communications shall be coordinated and controlled through the individual in charge of the ICS assisted by the senior official present for each employer.

Note to (q)(3)(i). The "senior official" at an emergency response is the most senior official on the site who has the responsibility for controlling the operation at the site. Initially it is the senior officer on the first-due piece of responding emergency apparatus to arrive on the incident scene. As more senior officers arrive (i.e., battalion chief, fire chief, state law enforcement official, site coordinator, etc.) the position is passed up the line of authority which has been previously established.

(q)(3)(ii) The individual in charge of the ICS shall identify, to the extent possible, all hazardous substances or conditions present and shall address as appropriate site analysis, use of engineering controls, maximum exposure limits, hazardous substance handling procedures, and use of any new technologies.

(q)(3)(iii) Based on the hazardous substances and/or conditions present, the individual in charge of the ICS shall implement appropriate emergency operations, and assure that the personal protective equipment worn is appropriate for the hazards to be encountered. However, personal protective equipment shall meet, at a minimum, the criteria contained in 29 CFR 1910.156(e) when worn while performing fire fighting operations beyond the incident stage for any incident.

[55 F.R. 14074, April 13, 1990]

(q)(3)(iv) Employees engaged in emergency response and exposed to hazardous substances presenting an inhalation hazard or potential inhalation hazard shall wear positive pressure self-contained breathing apparatus while engaged in emergency response, until such time that the individual in charge of the ICS determines through the use of air monitoring that a decreased level of respiratory protection will not result in hazardous exposures to employees.

(q)(3)(v) The individual in charge of the ICS

shall limit the number of emergency response personnel at the emergency site, in those areas of potential or actual exposure to incident or site hazards, to those who are actively performing emergency operations. However, operations in hazardous areas shall be performed using the buddy system in groups of two or more.

(q)(3)(vi) Back-up personnel shall stand by with equipment ready to provide assistance or rescue. Advance first aid support personnel, as a minimum, shall also stand by with medical equipment and transportation capability.

(q)(3)(vii) The individual in charge of the ICS shall designate a safety official, who is knowledgeable in the operations being implemented at the emergency response site, with specific responsibility to identify and evaluate hazards and to provide direction with respect to the safety of operations for the emergency at hand.

(q)(3)(viii) When activities are judged by the safety official to be an IDLH condition and/or to involve an imminent danger condition, the safety official shall have the authority to alter, suspend, or terminate those activities. The safety official shall immediately inform the individual in charge of the ICS of any actions needed to be taken to correct these hazards at the emergency scene.

(q)(3)(ix) After emergency operations have terminated, the individual in charge of the ICS shall implement appropriate decontamination procedures.

(q)(3)(x) When deemed necessary for meeting the tasks at hand, approved self-contained compressed air breathing apparatus may be used with approved cylinders from other approved self-contained compressed air breathing apparatus provided that such cylinders are of the same capacity and pressure rating. All compressed air cylinders used with self-contained breathing apparatus shall meet U.S. Department of Transportation and National Institute for Occupational Safety and Health criteria.

(q)(4) Skilled support personnel. Personnel, not necessarily an employer's own employees, who are skilled in the operation of certain equipment, such as mechanized earth moving or digging equipment or crane and hoisting equipment, and who are needed temporarily to perform immediate emergency support work that cannot reasonably be performed in a timely fashion by an employer's own employees, and who will be or may be exposed to the hazards at an emergency response scene, are not required to meet the training required in this paragraph for the employer's regular employees. However, these personnel shall be given an initial briefing at the site prior to their participation in any emergency response. The initial briefing shall include instruction in the wearing of appropriate personal protective equipment, what chemical hazards are involved, and what duties are to be performed. All other appropriate safety and health precautions provided to the employer's own employees shall be used to assure the safety and health of these personnel.

(q)(5) Specialist employees. Employees who, in the course of their regular job duties, work with and are trained in the hazards of specific hazardous substances, and who will be called upon to provide technical advice or assistance at a hazardous substance release incident to the individual in charge, shall receive training or demonstrate competency in the area of their specialization annually.

(q)(6) Training. Training shall be based on the duties and function to be performed by each responder of an emergency response organization. The skill and knowledge levels required for all new responders, those hired after the effective date of this standard, shall be conveyed to them through training before they are permitted to take part in actual emergency operations on an incident. Employees who participate, or are expected to participate, in emergency response shall be given training in accordance with the following paragraphs:

(q)(6)(i) First responder awareness level. First responders at the awareness level are individuals who are likely to witness or discover a hazardous substance release and who have been trained to initiate an emergency response sequence by notifying the proper authorities of the release. They would take no further action beyond notifying the authorities of the release. First responders at the awareness level shall have sufficient training or have had sufficient experience to objectively demonstrate competency in the following areas:

(q)(6)(i)(a) An understanding of what hazardous materials are, and the risks associated with them in an incident.

[55 F.R. 14074, April 13, 1990]

(q)(6)(i)(b) An understanding of the potential outcomes associated with an emergency created when hazardous substances are present.

[55 F.R. 14074, April 13, 1990]

(q)(6)(i)(c) The ability to recognize the presence of hazardous materials in an emergency.

[55 F.R. 14074, April 13, 1990]

(q)(6)(i)(d) The ability to identify the hazardous materials, if possible.

[55 F.R. 14074, April 13, 1990]

(q)(6)(i)(e) An understanding of the role of the first responder awareness individual in the employer's emergency response plan including the site security and control and the U.S. Department of Transportation's Emergency Response Guidebook.

(q)(6)(i)(f) The ability to realize the need for additional resources, and to make appropriate notifications to the communication center.

(q)(6)(ii) First responder operations level. First responders at the operations level are individuals who respond to releases or potential releases of hazardous substances as part of the initial response to the site for the purpose of protecting nearby persons, property, or the environment from the effects of the release. They are trained to respond in a defensive fashion without actually trying to stop the release. Their function is to contain the release from a safe distance, keep it from spreading, and prevent exposures. First responders at the operational level shall have received at least eight hours of training or have had sufficient experience to objectively demonstrate competency in the following areas in addition to those listed for the awareness level and the employer shall so certify:

(q)(6)(ii)(a) Knowledge of the basic hazard and risk assessment techniques.

(q)(6)(ii)(b) Know how to select and use proper personal protective equipment provided to the first responder operational level.

(q)(6)(ii)(c) An understanding of basic hazardous materials terms.

(q)(6)(ii)(d) Know how to perform basic control, containment and/or confinement operations within the capabilities of the resources and personal protective equipment available with their unit.

(q)(6)(ii)(e) Know how to implement basic decontamination procedures.

(q)(6)(ii)(f) An understanding of the relevant standard operating procedures and termination procedures.

(q)(6)(iii) Hazardous materials technician. Hazardous materials technicians are individuals who respond to releases or potential releases for the purpose of stopping the release. They assume a more aggressive role than a first responder at the operations level in that they will approach the point of release in order to plug, patch or otherwise stop the release of a hazardous substance. Hazardous materials technicians shall have received at least 24 hours of training equal to the first responder operations level and in addition have competency in the following areas and the employer shall so certify:

(q)(6)(iii)(a) Know how to implement the employer's emergency response plan.

(q)(6)(iii)(b) Know the classification, identification and verification of known and unknown materials by using field survey instruments and equipment.

(q)(6)(iii)(c) Be able to function within an assigned role in the Incident Command System.

(q)(6)(iii)(d) Know how to select and use proper specialized chemical personal protective equipment provided to the hazardous materials technician.

(q)(6)(iii)(e) Understand hazard and risk assessment techniques.

(q)(6)(iii)(f) Be able to perform advance control, containment, and/or confinement operations within the capabilities of the resources and personal protective equipment available with the unit.

(q)(6)(iii)(g) Understand and implement decontainment procedures.

(q)(6)(iii)(h) Understand termination procedures.

(q)(6)(iii)(i) Understand basic chemical and toxicological terminology and behavior.

(q)(6)(iv) Hazardous materials specialist. Hazardous materials specialists are individuals who respond with and provide support to hazardous materials technicians. Their duties parallel those of the hazardous materials technician; however, those duties require a more directed or specific knowledge of the various substances they may be called upon to contain. The hazardous materials specialist would also act as the site liaison with Federal, state, local, and other government authorities in regards to site activities. Hazardous materials specialists shall have received at least 24 hours of training equal to the technician level and in addition have competency in the following areas and the employer shall so certify:

(q)(6)(iv)(a) Know how to implement the local emergency response plan.

(q)(6)(iv)(b) Understand classification, identification and verification of known and unknown materials by using advanced survey instruments and equipment.

(q)(6)(iv)(c) Know of the state emergency response plan.

(q)(6)(iv)(d) Be able to select and use proper specialized chemical personal protective equipment provided to the hazardous materials specialist.

(q)(6)(iv)(e) Understand in-depth hazard and risk techniques.

(q)(6)(iv)(f) Be able to perform specialized control, containment, and/or confinement operations within the capabilities of the resources and personal protective equipment available.

(q)(6)(iv)(g) Be able to determine and implement decontamination procedures.

(q)(6)(iv)(h) Have the ability to develop a site safety and control plan.

(q)(6)(iv)(i) Understand chemical, radiological and toxicological terminology and behavior.

(q)(6)(v) On scene incident commander. Incident commanders, who will assume control of the incident scene beyond the first responder awareness level, shall receive at least 24 hours of training equal to the first responder operations level and in addition have competency in the following areas and the employer shall so certify:

(q)(6)(v)(a) Know and be able to implement the employer's incident command system.

(q)(6)(v)(b) Know how to implement the employer's emergency response plan.

(q)(6)(v)(c) Know and understand the hazards and risks associated with employees working in chemical protective clothing.

(q)(6)(v)(d) Know how to implement the local emergency response plan.

(q)(6)(v)(e) Know of the state emergency response plan and of the Federal Regional Response Team.

(q)(6)(v)(f) Know and understand the importance of decontamination procedures.

(q)(7) Trainers. Trainers who teach any of the above training subjects shall have satisfactorily completed a training course for teaching the subjects they are expected to teach, such as the course offered by the U.S. Fire Academy, or they shall have the training and/or academic credentials and instructional experience necessary to demonstrate competent instructional skills and a good command of the subject matter of the courses they are to teach.

[55 F.R. 14074, April 13, 1990]

(q)(8) Refresher training. (i) Those employees who are trained in accordance with paragraph (q)(6) of this section shall receive annual refresher training of sufficient content and duration to maintain their competencies, or shall demonstrate competency in those areas at least yearly.

(q)(8)(ii) A statement shall be made of the training or competency, and if a statement of competency is made, the employer shall keep a record of the methodology used to demonstrate competency.

(q)(9) Medical surveillance and consultation. (i) Members of an organized and designated HAZMAT team and hazardous materials specialists shall receive a baseline physical examination and be provided with medical surveillance as required in paragraph (f) of this section.

(q)(9)(ii) Any emergency response employees who exhibit signs or symptoms which may have resulted from exposure to hazardous substances during the course of an emergency incident, either immediately or subsequently, shall be provided with medical consultation as required in paragraph (f)(3)(ii) of this section.

(q)(10) Chemical protective clothing. Chemical protective clothing and equipment to be used by organized and designated HAZMAT team members, or to be used by hazardous materials specialists, shall meet the requirements of paragraphs (g)(3) through (5) of this section.

(q)(11) Post-emergency response operations. Upon completion of the emergency response, if it is determined that it is necessary to remove hazardous substances, health hazards, and materials contaminated with them (such as contaminated soil or other elements of the natural environment) from the site of the incident, the employer conducting the clean-up shall comply with one of the following:

(q)(11)(i) Meet all of the requirements of paragraphs (b) through (o) of this section; or

(q)(11)(ii) Where the clean-up is done on plant property using plant or workplace employees, such employees shall have completed the training requirements of the following: 29 CFR 1910.38(a); 1910.134; 1910.1200, and other appropriate safety and health training made necessary by the tasks that they are expected to perform such as personal protective equipment and decontamination procedures. All equipment to be used in the performance of the clean-up work shall be in serviceable condition and shall have been inspected prior to use.

APPENDICES TO 1910.120 HAZARDOUS WASTE OPERATIONS AND EMERGENCY RESPONSE

Note: The following appendices serve as non-mandatory guidelines to assist employees and employers in complying with the appropriate requirements of this section. However, paragraph 1910.120(g) makes mandatory in certain circumstances the use of Level A and Level B PPE protection.

Appendix A: Personal Protective Equipment Test Methods

This appendix sets forth the non-mandatory examples of tests which may be used to evaluate compliance with §1910.120(g)(4)(ii) and (iii). Other tests and challenge agents may be used to evaluate compliance.

A. TOTALLY-ENCAPSULATING CHEMICAL PROTECTIVE SUIT PRESSURE TEST

1.0 Scope

1.1 This practice measures the ability of a gas tight totally-encapsulating chemical protective suit material, seams, and closures to maintain a fixed positive pressure. The results of this practice allow the gas tight integrity of a totally-encapsulating chemical protective suit to be evaluated.

1.2 Resistance of the suit materials to permeation, penetration, and degradation by specific hazardous substances is not determined by this test method.

2.0 Definition of terms

2.1 "T*otally-encapsulating chemical protective suit (TECP suit)*" means a full body garment which is constructed of protective clothing materials; covers the wearer's torso, head, arms, legs and respirator; may cover the wearer's hands and feet with tightly attached gloves and boots; completely encloses the wearer and respirator by itself or in combination with the wearer's gloves and boots.

2.2 "*Protective clothing material*" means any

material or combination of materials used in an item of clothing for the purpose of isolating parts of the body from direct contact with a potentially hazardous liquid or gaseous chemicals.

2.3 "*Gas tight*" means, for the purpose of this test method, the limited flow of a gas under pressure from the inside of a TECP suit to atmosphere at a prescribed pressure and time interval.

3.0 Summary of test method

3.1 The TECP suit is visually inspected and modified for the test. The test apparatus is attached to the suit to permit inflation to the pre-test suit expansion pressure for removal of suit wrinkles and creases. The pressure is lowered to the test pressure and monitored for three minutes. If the pressure drop is excessive, the TECP suit fails the test and is removed from service. The test is repeated after leak location and repair.

4.0 Required supplies

4.1 Source of compressed air.

4.2 Test apparatus for suit testing, including a pressure measurement device with a sensitivity of at least 1/4 inch water gauge.

4.3 Vent valve closure plugs or sealing tape.

4.4 Soapy water solution and soft brush.

4.5 Stop watch or appropriate timing device.

5.0 Safety precautions

5.1 Care shall be taken to provide the correct pressure safety devices required for the source of compressed air used.

6.0 Test procedure

6.1 Prior to each test, the tester shall perform a visual inspection of the suit. Check the suit for seam integrity by visually examining the seams and gently pulling on the seams. Ensure that all air supply lines, fittings, visor, zippers, and valves are secure and show no signs of deterioration.

6.1.1 Seal off the vent valves along with any other normal inlet or exhaust points (such as umbilical air line fittings or facepiece opening) with tape or other appropriate means (caps, plugs, fixture, etc.). Care should be exercised in the sealing process not to damage any of the suit components.

6.1.2 Close all closure assemblies.

6.1.3 Prepare the suit for inflation by providing an improvised connection point on the suit for connecting an airline. Attach the pressure test apparatus to the suit to permit suit inflation from a compressed air source equipped with a pressure indicating regulator. The leak tightness of the pressure test apparatus should be tested before and after each test by closing off the end of the tubing attached to the suit and assuring a pressure of three inches water gauge for three minutes can be maintained. If a component is removed for the test, that component shall be replaced and a second test conducted with another component removed to permit a complete test of the ensemble.

6.1.4 The pre-test expansion pressure (A) and the suit test pressure (B) shall be supplied by the suit manufacturer, but in no case shall they be less than: (A) = three inches water gauge; and (B) = two inches water gauge. The ending suit pressure (C) shall be no less than 80 percent of the test pressure (B); i.e., the pressure drop shall not exceed 20 percent of the test pressure (B).

6.1.5 Inflate the suit until the pressure inside is equal to pressure (A), the pre-test expansion suit pressure. Allow at least one minute to fill out the wrinkles in the suit. Release sufficient air to reduce the suit pressure to pressure (B), the suit test pressure. Begin timing. At the end of three minutes, record the suit pressure as pressure (C), the ending suit pressure. The difference between the suit test pressure and the ending suit test pressure (B – C) shall be defined as the suit pressure drop.

6.1.6 If the suit pressure drop is more than 20 percent of the suit test pressure (B) during the three-minute test period, the suit fails the test and shall be removed from service.

7.0 Retest procedure

7.1 If the suit fails the test, check for leaks by inflating the suit to pressure (A) and brushing or wiping the entire suit (including seams, closures, lens gaskets, glove-to-sleeve joints, etc.) with a mild soap and water solution. Observe the suit for the formation of soap bubbles, which is an indication of a leak. Repair all identified leaks.

7.2 Retest the TECP suit as outlined in test procedure 6.0.

8.0 Report

8.1 Each TECP suit tested by this practice shall have the following information recorded:

8.1.1 Unique identification number, identifying brand name, date of purchase, material of construction, and unique fit features, e.g., special breathing apparatus.

8.1.2 The actual values for test pressures (A), (B), and (C) shall be recorded along with the specific observation times. If the ending pressure (C) is less than 80 percent of the test pressure (B), the suit shall be identified as failing the test. When possible, the specific leak location shall be identified in the test records. Retest pressure data shall be recorded as an additional test.

8.1.3 The source of the test apparatus used shall be identified and the sensitivity of the pressure gauge shall be recorded.

8.1.4 Records shall be kept for each pressure test even if repairs are being made at the test location.

Caution: Visually inspect all parts of the suit to be sure they are positioned correctly and secured tightly before putting the suit back into service. Special care should be taken to examine each exhaust valve to make sure it is not blocked.

Care should also be exercised to assure that the inside and outside of the suit is completely dry before it is put into storage.

B. TOTALLY-ENCAPSULATING CHEMICAL PROTECTIVE SUIT QUALITATIVE LEAK TEST

1.0 Scope

1.1 This practice semi-qualitatively tests gas tight totally-encapsulating chemical protective suit integrity by detecting inward leakage of ammonia vapor. Since no modifications are made to the suit to carry out this test, the results from this practice provide a realistic test for the integrity of the entire suit.

1.2 Resistance of the suit materials to permeation, penetration, and degradation is not determined by this test method. ASTM test methods are available to test suit materials for these characteristics and the tests are usually conducted by the manufacturers of the suits.

2.0 Definition of terms

2.1 "*Totally-encapsulating chemical protective suit (TECP suit)*" means a full body garment which is constructed of protective clothing materials; covers the wearer's torso, head, arms, legs and respirator; may cover the wearer's hands and feet with tightly attached gloves and boots; completely encloses the wearer and respirator by itself or in combination with the wearer's gloves and boots.

2.2 "*Protective clothing material*" means any material or combination of materials used in an item of clothing for the purpose of isolating parts of the body from direct contact with a potentially hazardous liquid or gaseous chemicals.

2.3 "*Gas tight*" means, for the purpose of this test method, the limited flow of a gas under pressure from the inside of a TECP suit to atmosphere at a prescribed pressure and time interval.

2.4 "*Intrusion coefficient*" means a number expressing the level of protection provided by a gas tight totally-encapsulating chemical protective suit. The intrusion coefficient is calculated by dividing the test room challenge agent concentration by the concentration of challenge agent found inside the suit. The accuracy of the intrusion coefficient is dependent on the challenge agent monitoring methods. The larger the intrusion coefficient, the greater the protection provided by the TECP suit.

3.0 Summary of recommended practice

3.1 The volume of concentrated aqueous ammonia solution (ammonium hydroxide NH_4OH) required to generate the test atmosphere is determined using the directions outlined in 6.1.1. The suit is donned by a person wearing the appropriate respiratory equipment (either a positive pressure self-contained breathing apparatus or a positive pressure supplied air respirator) and worn inside the enclosed test room. The concentrated aqueous ammonia solution is taken by the suited individual into the test room and poured into an open plastic pan. A two-minute evaporation period is observed before the test room concentration is measured, using a high range ammonia length of stain detector tube. When the ammonia vapor reaches a concentration of between 1000 and 1200 ppm, the suited individual starts a standardized exercise protocol to stress and flex the suit. After this protocol is completed, the test room concentration is measured again. The suited individual exits the test room and his stand-by person measures the ammonia concentration inside the suit using a low range ammonia length of stain detector tube or other more sensitive ammonia detector. A stand-by person is required to observe the test individual during the test procedure; aid the person in donning and doffing the TECP suit; and monitor the suit interior. The intrusion coefficient of the suit can be calculated by dividing the average test area concentration by the interior suit concentration. A colorimetric ammonia indicator strip of bromophenol blue or equivalent is placed on the inside of the suit facepiece lens so that the suited individual is able to detect a color change and know if the suit has a significant leak. If a color change is observed the individual shall leave the test room immediately.

4.0 Required supplies

4.1 A supply of concentrated aqueous ammonium hydroxide (58 percent by weight).

[55 F.R. 14074, April 13, 1990]

4.2 A supply of bromophenol blue indicating paper or equivalent, sensitive to 5–10 ppm ammonia or greater over a two-minute period of exposure [pH 3.0 (yellow) to pH 4.6 (blue)].

4.3 A supply of high range (0.5–10 volume

percent) and low range (5–700 ppm) detector tubes for ammonia and the corresponding sampling pump. More sensitive ammonia detectors can be substituted for the low range detector tubes to improve the sensitivity of this practice.

4.4 A shallow plastic pan (PVC) at least 12″:14″:1″ and a half pint plastic container (PVC) with tightly closing lid.

4.5 A graduated cylinder or other volumetric measuring device of at least 50 milliliters in volume with an accuracy of at least ±1 milliliters.

5.0 Safety precautions

5.1 Concentrated aqueous ammonium hydroxide, NH_4OH, is a corrosive volatile liquid requiring eye, skin, and respiratory protection. The person conducting the test shall review the MSDS for aqueous ammonia.

5.2 Since the established permissible exposure limit for ammonia is 50 ppm as a 15-minute STEL, only persons wearing a positive pressure self-contained breathing apparatus or a positive pressure supplied air respirator shall be in the chamber. Normally, only the person wearing the totally-encapsulating suit will be inside the chamber. A stand-by person shall have a positive pressure self-contained breathing apparatus or a positive pressure supplied air respirator available to enter the test area, should the suited individual need assistance.

[55 F.R. 14074, April 13, 1990]

5.3 A method to monitor the suited individual must be used during this test. Visual contact is the simplest but other methods using communication devices are acceptable.

5.4 The test room shall be large enough to allow the exercise protocol to be carried out and then to be ventilated to allow for easy exhaust of the ammonia test atmosphere after the test(s) are completed.

5.5 Individuals shall be medically screened for the use of respiratory protection and checked for allergies to ammonia before participating in this test procedure.

6.0 Test procedure

6.1.1 Measure the test area to the nearest foot and calculate its volume in cubic feet. Multiply the test area volume by 0.2 milliliters of concentrated aqueous ammonia solution per cubic foot of test area volume to determine the approximate volume of concentrated aqueous ammonia required to generate 1000 ppm in the test area.

6.1.2 Measure this volume from the supply of concentrated aqueous ammonia and place it into a closed plastic container.

6.1.3 Place the container, several high range ammonia detector tubes, and the pump in the clean test pan and position it near the test area entry door so that the suited individual has easy access to these supplies.

6.2.1 In a non-contaminated atmosphere, open a pre-sealed ammonia indicator strip and fasten one end of the strip to the inside of the suit face shield lens where it can be seen by the wearer. Moisten the indicator strip with distilled water. Care shall be taken not to contaminate the detector part of the indicator paper by touching it. A small piece of masking tape or equivalent should be used to attach the indicator strip to the interior of the suit face shield.

6.2.2 If problems are encountered with this method of attachment, the indicator strip can be attached to the outside of the respirator facepiece lens being used during the test.

6.3 Don the respiratory protective device normally used with the suit, and then don the

TECP suit to be tested. Check to be sure all openings which are intended to be sealed (zippers, gloves, etc.) are completely sealed. DO NOT, however, plug off any venting valves.

6.4 Step into the enclosed test room such as a closet, bathroom, or test booth, equipped with an exhaust fan. No air should be exhausted from the chamber during the test because this will dilute the ammonia challenge concentrations.

6.5 Open the container with the premeasured volume of concentrated aqueous ammonia within the enclosed test room, and pour the liquid into the empty plastic test pan. Wait two minutes to allow for adequate volatilization of the concentrated aqueous ammonia. A small mixing fan can be used near the evaporation pan to increase the evaporation rate of the ammonia solution.

6.6 After two minutes a determination of the ammonia concentration within the chamber should be made using the high range colorimetric detector tube. A concentration of 1000 ppm ammonia or greater shall be generated before the exercises are started.

6.7 To test the integrity of the suit, the following four-minute exercise protocol should be followed:

6.7.1 Raising the arms above the head with at least 15 raising motions completed in one minute.

6.7.2 Walking in place for one minute with at least 15 raising motions of each leg in a one-minute period.

6.7.3 Touching the toes with at least 10 complete motions of the arms from above the head to touching of the toes in a one-minute period.

6.7.4 Knee bends with at least 10 complete standing and squatting motions in a one-minute period.

6.8 If at any time during the test the colorimetric indicating paper should change colors, the test should be stopped and section 6.10 and 6.12 initiated (see paragraph 4.2).

6.9 After completion of the test exercise, the test area concentration should be measured again using the high range colorimetric detector tube.

6.10 Exit the test area.

6.11 The opening created by the suit zipper or other appropriate suit penetration should be used to determine the ammonia concentration in the suit with the low range length of stain detector tube or other ammonia monitor. The internal TECP suit air should be sampled far enough from the enclosed test area to prevent a false ammonia reading.

6.12 After completion of the measurement of the suit interior ammonia concentration, the test is concluded and the suit is doffed and the respirator removed.

6.13 The ventilating fan for the test room should be turned on and allowed to run for enough time to remove the ammonia gas. The fan shall be vented to the outside of the building.

6.14 Any detectable ammonia in the suit interior (five ppm ammonia (NH_3) or more for the length of stain detector tube) indicates that the suit has failed the test. When other ammonia detectors are used a lower level of detection is possible, and it should be specified as the pass/fail criteria.

6.15 By following this test method, an intrusion coefficient of approximately 200 or more can be measured with the suit in a completely operational condition. If the intrusion coefficient is 200 or more, then the suit is suitable for emergency response and field use.

7.0 Retest procedures

7.1 If the suit fails this test, check for leaks by following the pressure test in test A above.

7.2 Retest the TECP suit as outlined in the test procedure 6.0.

8.0 Report

8.1 Each gas tight totally-encapsulating chemical protective suit tested by this practice shall have the following information recorded.

8.1.1 Unique identification number, identifying brand name, date of purchase, material of construction, and unique suit features, e.g., special breathing apparatus.

8.1.2 General description of test room used for the test.

8.1.3 Brand name and purchase date of ammonia detector strips and color change data.

8.1.4 Brand name, sampling range, and expiration date of the length of stain ammonia detector tubes. The brand name and model of the sampling pump should also be recorded. If another type of ammonia detector is used, it should be identified along with its minimum detection limit for ammonia.

8.1.5 Actual test results shall list the two test area concentrations, their average, the interior suit concentration, and the calculated intrusion coefficient. Retest data shall be recorded as an additional test.

8.2 The evaluation of the data shall be specified as "suit passed" or "suit failed," and the date of the test. Any detectable ammonia (five ppm or greater for the length of stain detector tube) in suit interior indicates the suit has failed this test. When other ammonia detectors are used, a lower level of detection is possible and it should be specified as the pass/fail criteria.

Caution: Visually inspect all parts of the suit to be sure they are positioned correctly and secured tightly before putting the suit back into service. Special care should be taken to examine each exhaust valve to make sure it is not blocked.

Care should also be exercised to assure that the inside and outside of the suit is completely dry before it is put into storage.

Appendix B: General Description and Discussion of the Levels of Protection and Protective Gear

This appendix sets forth information about personal protective equipment (PPE) protection levels which may be used to assist employers in complying with the PPE requirements of this section.

As required by the standard, PPE must be selected which will protect employees from the specific hazards which they are likely to encounter during their work on-site.

Selection of the appropriate PPE is a complex process which should take into consideration a variety of factors. Key factors involved in this process are identification of the hazards, or suspected hazards; their routes of potential hazard to employees (inhalation, skin absorption, ingestion, and eye or skin contact); and the performance of the PPE *materials* (and seams) in providing a barrier to these hazards. The amount of protection provided by PPE is material-hazard specific.

That is, protective equipment materials will protect well against some hazardous substances and poorly, or not at all, against others. In many instances, protective equipment materials cannot be found which will provide continuous protection from the particular hazardous substance. In these cases, the breakthrough time of the protective material should exceed the work durations.

[55 F.R. 14074, April 13, 1990]

Other factors in this selection process to be considered are matching the PPE to the employee's work requirements and task-specific conditions. The durability of PPE materials, such as tear strength and seam strength, should be considered in relation to the employee's tasks. The effects of PPE in relation to heat stress and task duration are a factor in selecting and using PPE. In some cases layers of PPE may be necessary to provide sufficient protection, or to protect expensive PPE inner garments, suits or equipment.

The more that is known about the hazards at the site, the easier the job of PPE selection becomes. As more information about the hazards and conditions at the site becomes available, the site supervisor can make decisions to up-grade or down-grade the level of PPE protection to match the tasks at hand.

The following are guidelines which an employer can use to begin the selection of the appropriate PPE. As noted above, the site information may suggest the use of combinations of PPE selected from the different protection levels (i.e., A, B, C, or D) as being more suitable to the hazards of the work. It should be cautioned that the listing below does not fully address the performance of the specific PPE material in relation to the specific hazards at the job site, and that PPE selection, evaluation and re-selection is an ongoing process until sufficient information about the hazards and PPE performance is obtained.

PART A. Personal protective equipment is divided into four categories based on the degree of protection afforded. (See Part B of this appendix for further explanation of Levels A, B, C, and D hazards.)

I. Level A. To be selected when the greatest level of skin, respiratory, and eye protection is required.

The following constitute Level A equipment; it may be used as appropriate.

1. Positive pressure, full-facepiece self-contained breathing apparatus (SCBA), or positive pressure supplied air respirator with escape SCBA, approved by the National Institute for Occupational Safety and Health (NIOSH).
2. Totally-encapsulating chemical-protective suit.
3. Coveralls (optional, as applicable).
4. Long underwear (optional, as applicable).
5. Gloves, outer, chemical-resistant.
6. Gloves, inner, chemical-resistant.
7. Boots, chemical-resistant, steel toe and shank.
8. Hard hat (under suit) (optional, as applicable).
9. Disposable protective suit, gloves, and boots (depending on suit construction, may be worn over totally-encapsulating suit).

II. Level B. The highest level of respiratory protection is necessary but a lesser level of skin protection is needed.

The following constitute Level B equipment; it may be used as appropriate.

1. Positive pressure, full-facepiece self-contained breathing apparatus (SCBA),

or positive pressure supplied air respirator with escape SCBA (NIOSH approved).

2. Hooded chemical-resistant clothing (overalls and long-sleeved jacket; coveralls; one or two-piece chemical-splash suit; disposable chemical-resistant overalls).
3. Coveralls (optional, as applicable).
4. Gloves, outer, chemical-resistant.
5. Gloves, inner, chemical-resistant.
6. Boots, outer, chemical-resistant, steel toe and shank.
7. Boot-covers, outer, chemical-resistant (disposable) (optional, as applicable).
8. Hard hat (optional, as applicable).
9. [Reserved]
10. Face shield (optional, as applicable).

III. Level C. The concentration(s) and type(s) of airborne substance(s) is known and the criteria for using air purifying respirators are met.

The following constitute Level C equipment; it may be used as appropriate.

1. Full-face or half-mask air purifying respirators (NIOSH approved).
2. Hooded chemical-resistant clothing (overalls; two-piece chemical-splash suit; disposable chemical-resistant overalls).
3. Coveralls (optional, as applicable).
4. Gloves, outer, chemical-resistant.
5. Gloves, inner, chemical-resistant.
6. Boots, outer, chemical-resistant, steel toe and shank (optional, as applicable).
7. Boot-covers, outer, chemical-resistant (disposable) (optional, as applicable).
8. Hard hat (optional, as applicable).
9. Escape mask (optional, as applicable).
10. Face shield (optional, as applicable).

IV. Level D. A work uniform affording minimal protection, used for nuisance contamination only.

The following constitute Level D equipment; it may be used as appropriate.

1. Coveralls.
2. Gloves (optional, as applicable).
3. Boots/shoes, chemical-resistant, steel toe and shank.
4. Boots, outer, chemical-resistant (disposable) (optional, as applicable).
5. Safety glasses or chemical splash goggles.
6. Hard hat (optional, as applicable).
7. Escape mask (optional, as applicable).
8. Face shield (optional, as applicable).

PART B. The types of hazards for which levels A, B, C, and D protection are appropriate are described below:

I. Level A. Level A protection should be used when:

1. The hazardous substance has been identified and requires the highest level of protection for skin, eyes, and the respiratory system based on either the measured (or potential for) high concentration of atmospheric vapors, gases, or particulates; or the site operations and work functions involve a high potential for splash, immersion, or exposure to unexpected vapors, gases, or particulates of materials that are harmful to skin or capable of being absorbed through the skin;
2. Substances with a high degree of hazard to the skin are known or suspected to

be present, and skin contact is possible; or

3. Operations are being conducted in confined, poorly ventilated areas, and the absence of conditions requiring Level A have not yet been determined.

II. Level B. Level B protection should be used when:

1. The type and atmospheric concentration of substances have been identified and require a high level of respiratory protection, but less skin protection;
2. The atmosphere contains less than 19.5 percent oxygen; or
3. The presence of incompletely identified vapors or gases is indicated by a direct-reading organic vapor detection instrument, but vapors and gases are not suspected of containing high levels of chemicals harmful to skin or capable of being absorbed through the skin.

Note: This involves atmospheres with IDLH concentrations of specific substances that present severe inhalation hazards and that do not represent a severe skin hazard; or that do not meet the criteria for use of air-purifying respirators.

III. Level C. Level C protection should be used when:

1. The atmospheric contaminants, liquid splashes, or other direct contact will not adversely affect or be absorbed through any exposed skin;
2. The types of air contaminants have been identified, concentrations measured, and an air-purifying respirator is available that can remove the contaminants; and
3. All criteria for the use of air-purifying respirators are met.

IV. Level D. Level D protection should be used when:

1. The atmosphere contains no known hazard; and
2. Work functions preclude splashes, immersion, or the potential for unexpected inhalation of or contact with hazardous levels of any chemicals.

Note: As stated in NFPA 1992—Standard on Liquid Splash-Protective Suits for Hazardous Chemical Emergencies (EPA Level B Protective Clothing).

[55 F.R. 14074, April 13, 1990]

As an aid in selecting suitable chemical protective clothing, it should be noted that the National Fire Protection Association is developing standards on chemical protective clothing. These standards are currently undergoing public review prior to adoption, including:

- NFPA 1991: Standard on Vapor-Protective Suits for Hazardous Chemical Emergencies (EPA Level A Protective Clothing)
- NFPA 1991: Standard on Liquid Splash-Protective Suits for Hazardous Chemical Emergencies (EPA Level B Protective Clothing)
- NFPA 1993: Standard on Liquid Splash-Protective Suits for Non-emergency, Non-flammable Hazardous Chemical Situations (EPA Level B Protective Clothing)

These standards would apply documentation and performance requirements to the manufacture of chemical protective suits. Chemical protective suits meeting these requirements would be labelled as compliant with the appropriate standard. When these standards are adopted by the National Fire Protection Association, it is recommended that chemical protective suits which meet these standards be used.

Appendix C: Compliance Guidelines

1. Occupational safety and health program. Each hazardous waste site clean-up effort will require an occupational safety and health program headed by the site coordinator or the employer's representative. The purpose of the program will be the protection of employees at the site and will be an extension of the employer's overall safety and health program. The program will need to be developed before work begins on the site and implemented as work proceeds as stated in paragraph (b). The program is to facilitate coordination and communication of safety and health issues among personnel responsible for the various activities which will take place at the site. It will provide the overall means for planning and implementing the needed safety and health training and job orientation of employees who will be working at the site. The program will provide the means for identifying and controlling worksite hazards and the means for monitoring program effectiveness. The program will need to cover the responsibilities and authority of the site coordinator or the employer's manager on the site for the safety and health of employees at the site, and the relationships with contractors or support services as to what each employer's safety and health responsibilities are for their employees on the site. Each contractor on the site needs to have its own safety and health program so structured that it will smoothly interface with the program of the site coordinator or principal contractor.

Also those employers involved with treating, storing or disposal of hazardous waste as covered in paragraph (p) must have implemented a safety and health program for their employees. This program is to include the hazard communication program required in paragraph (p)(1) and the training required in paragraphs (p)(7) and (p)(8) as parts of the employer's comprehensive overall safety and health program. This program is to be in writing.

Each site or workplace safety and health program will need to include the following: (1) Policy statements of the line of authority and accountability for implementing the program, the objectives of the program and the role of the site safety and health supervisor or manager and staff; (2) means or methods for the development of procedures for identifying and controlling workplace hazards at the site; (3) means or methods for the development and communication to employees of the various plans, work rules, standard operating procedures and practices that pertain to individual employees and supervisors; (4) means for the training of supervisors and employees to develop the needed skills and knowledge to perform their work in a safe and healthful manner; (5) means to anticipate and prepare for emergency situations; and (6) means for obtaining information feedback to aid in evaluating the program and for improving the effectiveness of the program. The management and employees should be trying continually to improve the effectiveness of the program, thereby enhancing the protection being afforded those working on the site.

Accidents on the site or workplace should be investigated to provide information on how such occurrences can be avoided in the future. When injuries or illnesses occur on the site or workplace, they will need to be investigated to determine what needs to be done to prevent this incident from occurring again. Such information will need to be used as feedback on the effectiveness of the program and the information turned into positive steps to prevent any reoccurrence. Receipt of employee suggestions or complaints relating to safety and health issues involved with site or workplace activities is also a feedback mechanism that can be used effectively to improve the program and may serve in part as an evaluative tool(s).

For the development and implementation of the program to be the most effective, professional safety and health personnel should be used. Certified Safety Professionals, Board Certified Industrial Hygienists or Registered Professional Safety Engineers are good examples of professional stature for safety and health managers who will administer the employer's program.

2. Training. The training programs for employees subject to the requirements of paragraph (e) of this standard should address: the safety and health hazards employees should expect to find on hazardous waste clean-up sites; what control measures or techniques are effective for those hazards; what monitoring procedures are effective in characterizing exposure levels; what makes an effective employer's safety and health program; what a site safety and health plan should include; hands-on training with personal protective equipment and clothing they may be expected to use; the contents of the OSHA standard relevant to the employee's duties and function; and employee's responsibilities under OSHA and other regulations. Supervisors will need training in their responsibilities under the safety and health program and its subject areas such as the spill containment program, the personal protective equipment program, the medical surveillance program, the emergency response plan and other areas.

The training programs for employees subject to the requirements of paragraph (p) of this standard should address: the employer's safety and health program elements impacting employees; the hazard communication program; the medical surveillance program; the hazards and the controls for such hazards that employees need to know for their job duties and functions. All require annual refresher training.

The training programs for employees covered by the requirements of paragraph (q) of this standard should address those competencies required for the various levels of response such as: the hazards associated with hazardous substances; hazard identification and awareness; notification of appropriate persons; the need for and use of personal protective equipment including respirators; the decontamination procedures to be used; preplanning activities for hazardous substance incidents including the emergency response plan; company standard operating procedures for hazardous substance emergency responses; the use of the incident command system and other subjects. Hands-on training should be stressed whenever possible. Critiques done after an incident which include an evaluation of what worked and what did not and how could the incident be better handled the next time may be counted as training time.

For hazardous materials specialists (usually members of hazardous materials teams), the training should address the care, use and/or testing of chemical protective clothing including totally-encapsulating suits, the medical surveillance program, the standard operating procedures for the hazardous materials team including the use of plugging and patching equipment and other subject areas.

Officers and leaders who may be expected to be in charge at an incident should be fully knowledgeable of their company's incident command system. They should know where and how to obtain additional assistance and be familiar with the local district's emergency response plan and the state emergency response plan.

Specialist employees such as technical experts, medical experts or environmental experts that work with hazardous materials in their regular jobs, who may be sent to the incident scene by the shipper, manufacturer or governmental agency to advise and assist the person in charge of the incident should have training on an annual basis. Their training should include the care and use of personal protective equipment including respirators; knowledge of the incident command system and how they are to relate to it; and those areas needed to keep them current in their respective field as it relates to safety and health involving specific hazardous substances.

Those skilled support personnel, such as employees who work for public works departments or equipment operators who operate bulldozers, sand trucks, backhoes, etc., who may be called to the incident scene to provide emergency support assistance, should have at least a safety and health briefing before entering the area of potential or actual exposure. These skilled support personnel, who have not been a part of the emergency response plan and do not meet the training requirements, should be made aware of the hazards they face and should be provided all necessary protective clothing and equipment required for their tasks.

There are two National Fire Protection Association standards, NFPA 472—"Standard for Professional Competence of Responders to Hazardous Material Incidents" and NFPA 471—"Recommended Practice for Responding to Hazardous Material Incidents," which are excellent resource documents to aid fire departments and other emergency response organizations in developing their training program materials. NFPA 472 provides guidance on the skills and knowledge needed for first responder awareness level, first responder operations level, hazmat technicians, and hazmat specialist. It also offers guidance for the officer corp who will be in charge of hazardous substance incidents.

[55 F.R. 14074, April 13, 1990]

3. Decontamination. Decontamination procedures should be tailored to the specific hazards of the site, and may vary in complexity and number of steps, depending on the level of hazard and the employee's exposure to the hazard. Decontamination procedures and PPE decontamination methods will vary depending upon the specific substance, since one procedure or method may not work for all substances. Evaluation of decontamination methods and procedures should be performed, as necessary, to assure that employees are not exposed to hazards by re-using PPE. References in Appendix D may be used for guidance in establishing an effective decontamination program. In addition, the U.S. Coast Guard's manual, "Policy Guidance for Response to Hazardous Chemical Releases," U.S. Department of Transportation, Washington, DC (COMDTINST M16465.30) is a good reference for establishing an effective decontamination program.

4. Emergency response plans. States, along with designated districts within the states, will be developing or have developed local emergency response plans. These state and district plans should be utilized in the emergency response plans called for in the standard. Each employer should assure that its emergency response plan is compatible with the local plan. The major reference being used to aid in developing the state and local district plans is the *Hazardous Materials Emergency Planning Guide,* NRT-1. The current Emergency Response Guidebook from the U.S. Department of Transportation, CMA's CHEMTREC and the Fire Service Emergency Management Handbook may also be used as resources.

Employers involved with treatment, storage, and disposal facilities for hazardous waste, which have the required contingency plan called for by their permit, would not need to duplicate the same planning elements. Those items of the emergency response plan that are properly addressed in the contingency plan may be substituted into the emergency response plan required in 1910.120 or otherwise kept together for employer and employee use.

5. Personal protective equipment programs. The purpose of personal protective clothing and equipment (PPE) is to shield or isolate individuals from the chemical, physical, and biologic hazards that may be encountered at a hazardous substance site.

As discussed in Appendix B, no single combination of protective equipment and clothing is capable of protecting against all hazards. Thus, PPE should be used in conjunction with other protective methods and its effectiveness evaluated periodically.

The use of PPE can itself create significant worker hazards, such as heat stress; physical and psychological stress; and impaired vision, mobility, and communication. For any given situation, equipment and clothing should be selected that provide an adequate level of protection. However, overprotection, as well as under-protection, can be hazardous and should be avoided where possible.

Two basic objectives of any PPE program should be to protect the wearer from safety and health hazards, and to prevent injury to the wearer from incorrect use and/or malfunction of the PPE. To accomplish these goals, a comprehensive PPE program should include hazard identification, medical monitoring, environmental surveillance, selection, use, maintenance, and decontamination of PPE and its associated training.

The written PPE program should include policy statements, procedures, and guidelines. Copies should be made available to all employees, and a reference copy should be made available at the worksite. Technical data on equipment, maintenance manuals, relevant regulations, and other essential information should also be collected and maintained.

6. Incident command system (ICS). Paragraph 1910.120(q)(3)(ii) requires the implementation of an ICS. The ICS is an organized approach to effectively control and manage operations at an emergency incident. The individual in charge of the ICS is the senior official responding to the incident. The ICS is not much different than the "command post" approach used for many years by the fire service. During large complex fires involving several companies and many pieces of apparatus, a command post would be established. This enabled one individual to be in charge of managing the incident, rather than having several officers from different companies making separate, and sometimes conflicting, decisions. The individual in charge of the command post would delegate responsibility for performing various tasks to subordinate officers. Additionally, all communications were routed through the command post to reduce the number of radio transmissions and eliminate confusion. However, strategy, tactics, and all decisions were made by one individual.

The ICS is a very similar system, except it is implemented for emergency response to all incidents, both large and small, that involve hazardous substances.

For a small incident, the individual in charge of the ICS may perform many tasks of the ICS. There may not be any, or little, delegation of tasks to subordinates. For example, in response to a small incident, the individual in charge of the ICS, in addition to normal command activities, may become the safety officer and may designate only one employee (with proper equipment) as a back-up to provide assistance if needed. OSHA does recommend, however, that at least two employees be designated as back-up personnel since the assistance needed may include rescue.

To illustrate the operation of the ICS, the following scenario might develop during a small incident, such as an overturned tank truck with a small leak of flammable liquid.

The first responding senior officer would implement and take command of the ICS. That person would size-up the incident and determine if additional personnel and apparatus were necessary; would determine what actions to take to control the leak; and determine the proper level of personal protective equipment. If additional assistance is not needed, the individual in charge of the ICS would implement actions to stop and control the leak using the fewest number of personnel that can effectively accomplish the tasks. The individual in charge of the ICS then would designate himself as the safety officer and two other employees as a back-up in case rescue may become necessary. In this scenario, decontamination procedures would not be necessary.

A large complex incident may require many employees and difficult, time-consuming efforts to control. In these situations, the individual in charge of the ICS will want to delegate different tasks to subordinates in order to maintain a span of control that will keep the number of subordinates that are reporting to a manageable level.

Delegation of task at large incidents may be by location, where the incident scene is divided into sectors, and subordinate officers coordinate activities within the sector that they have been assigned.

Delegation of tasks can also be by function. Some of the functions that the individual in charge of the ICS may want to delegate at a large incident are: medical services; evacuation; water supply; resources (equipment, apparatus); media relations; safety; and site control (integrate activities with policy for crowd and traffic control). Also for a large incident, the individual in charge of the ICS will designate several employees as back-up personnel, and a number of safety officers to monitor conditions and recommend safety precautions.

Therefore, no matter what size or complexity an incident may be, by implementing an ICS there will be *one individual in charge* who makes the decisions and gives directions; and all actions and communications are coordinated through one central point of command. Such a system should reduce confusion, improve safety, organize and coordinate actions, and should facilitate effective management of the incident.

7. Site safety and control plans. The safety and security of response personnel and others in the area of emergency response incident site should be of primary concern to the incident commander. The use of a site safety and control plan could greatly assist those in charge of assuring the safety and health of employees on the site.

A comprehensive site safety and control plan should include the following: summary analysis of hazards on the site and a risk analysis of those hazards; site map or sketch; site work zones (clean zone, transition or decontamination zone, work or hot zone); use of the buddy system; site communications; command post or command center; standard operating procedures and safe work practices; medical assistance and triage area; hazard monitoring plan (air contaminate monitoring, etc.); decontamination procedures and area; and other relevant areas. This plan should be a part of the employer's emergency response plan or an extension of it to the specific site.

8. Medical surveillance programs. Workers handling hazardous substances may be exposed to toxic chemicals, safety hazards, biologic hazards, and radiation. Therefore, a medical surveillance program is essential to assess and monitor workers' health and fitness for employment in hazardous waste operations and during the course of work; to provide emergency and other treatment as needed; and to keep accurate records for future reference.

9. New technology and spill containment programs. Where hazardous substances may be released by spilling from a container that will expose employees to the hazards of the materials, the employer will need to implement a program to contain and control the spilled material. Diking and ditching, as well as use of absorbents like diatomaceous earth, are traditional techniques which have proven to be effective over the years, However, in recent years new products have come into the marketplace, the use of which complement and increase the effectiveness of these traditional methods. These new products also provide emergency responders and others with additional tools or agents to use to reduce the hazards of spilled materials.

These agents can be rapidly applied over a large area and can be uniformly applied or otherwise can be used to build a small dam, thus improving the workers' ability to control spilled material. These application techniques enhance the intimate contact between the agent and spilled material, allowing for the quickest effect by the agent or quickest control of the spilled material. Agents are available to solidify liquid spilled materials, to suppress vapor generation from spilled materials, and to do both. Some special agents, which when applied as recommended by the manufacturer, will react in a controlled manner with the spilled material to neutralize acids or caustics, or greatly reduce the level of hazard of the spilled material.

There are several modern methods and devices for use by emergency response personnel or others involved with spill control efforts to safely apply spill control agents to control spilled material hazards. These include portable pressurized applicators similar to hand-held portable fire-extinguishing devices, and nozzle and hose systems similar to portable fire fighting foam systems which allow the operator to apply the agent without having to come into contact with the spilled material. The operator is able to apply the agent to the spilled material from a remote position.

The solidification of liquids provides for rapid containment and isolation of hazardous substance spills. By directing the agent at run-off points or at the edges of the spill, the reactant solid will automatically create a barrier to slow or stop the spread of the material. Clean-up of hazardous substances is greatly improved when solidifying agents, acid or caustic neutralizers, or activated carbon absorbents are used. Properly applied, these agents can totally solidify liquid hazardous substances or neutralize or absorb them, which results in materials which are less hazardous and easier to handle, transport, and dispose of. The concept of spill treatment to create less hazardous substances will improve the safety and level of protection of employees working at spill clean-up operations or emergency response operations to spills of hazardous substances.

The use of vapor suppression agents for volatile hazardous substances, such as flammable liquids and those substances which present an inhalation hazard, is important for protecting workers. The rapid and uniform distribution of the agent over the surface of the spilled material can provide quick vapor knockdown. There are temporary and long-term foam-type agents which are effective on vapors and dusts, and activated carbon adsorption agents which are effective for vapor control and soaking-up of the liquid. The proper use of hose lines or hand-held portable pressurized applicators provides good mobility and permits the worker to deliver the agent from a safe distance without having to step into the untreated spilled material. Some of these systems can be recharged in the field to provide coverage of larger spill areas than the design limits of a single charged applicator unit. Some of the more effective agents can solidify the liquid flammable hazardous substances and at the same time elevate the flash point above 140°F so the resulting substance may be handled as a nonhazardous waste material if it meets the U.S. Environmental Protection Agency's 40 CFR Part 261 requirements. (See particularly paragraph 261.21.)

All workers performing hazardous substance spill control are expected to wear the proper protective clothing and equipment for the materials present and to follow the employer's established standard operating procedures for spill control. All involved workers need to be trained in the established standard operating procedures; in the use and care of spill control equipment; and in the associated hazards and control of such hazards of spill containment work.

These new tools and agents are the things that employers will want to evaluate as part of their new technology program. The treatment of spills of hazardous substances or wastes at an emergency incident as part of the immediate spill containment and control efforts is sometimes acceptable to EPA, and a permit exception is described in 40 CFR 264.1(g)(8) and 265.1(c)(11).

[55 F.R. 14074, April 13, 1990]

The *Occupational Safety and Health Guidance Manual for Hazardous Waste Site Activities,* developed by the National Institute for Occupational Safety and Health (NIOSH), the Occupational Safety and Health Administration (OSHA), the U.S. Coast Guard (USCG), and the Environmental Protection Agency (EPA), October 1985 provides an excellent example of the types of medical testing that should be done as part of a medical surveillance program.

Appendix D: References

The following references may be consulted for further information on the subject of this standard:

1. OSHA Instruction DFO CPL 2.70—January 29, 1986, *Special Emphasis Program: Hazardous Waste Sites.*

2. OSHA Instruction DFO CPL 2-2.37A—January 29, 1986, *Technical Assistance and Guidelines for Superfund and Other Hazardous Waste Site Activities.*

3. OSHA Instructions DTS CPL 2.74—January 29, 1986, *Hazardous Waste Activity Form, OSHA 175.*

4. *Hazardous Waste Inspections Reference Manual,* U.S. Department of Labor, Occupational Safety and Health Administration, 1986.

5. Memorandum of Understanding Among the National Institute for Occupational Safety and Health, the Occupational Safety and Health Administration, the United States Coast Guard, and the United States Environmental Protection Agency, *Guidance for Worker Protection During Hazardous Waste Site Investigations and Clean-up and Hazardous Substance Emergencies.* December 18, 1980.

6. *National Priorities List,* 1st Edition, October 1984; U.S. Environmental Protection Agency, Revised periodically.

7. *The Decontamination of Response Personnel,* Field Standard Operating Procedures (F.S.O.P.) 7; U.S. Environmental Protection Agency, Office of Emergency and Remedial Response, Hazardous Response Support Division, December 1984.

8. *Preparation of a Site Safety Plan,* Field Standard Operating Procedures (F.S.O.P.) 9; U.S. Environmental Protection Agency, Office of Emergency and Remedial Response, Hazardous Response Support Division, April 1985.

9. *Standard Operating Safety Guidelines;* U.S. Environmental Protection Agency, Office of Emergency and Remedial Response, Hazardous Response Support Division, November 1984.

10. *Occupational Safety and Health Guidance Manual for Hazardous Waste Site Activities,* National Institute for Occupational Safety and Health (NIOSH), Occupational Safety and Health Administration (OSHA), U.S. Coast Guard (USCG), and Environmental Protection Agency (EPA); October 1985.

11. *Protecting Health and Safety at Hazardous Waste Sites: An Overview,* U.S. Environmental Protection Agency, EPA/625/9-85/006; September 1985.

12. *Hazardous Waste Sites and Hazardous Substance Emergencies,* NIOSH Worker Bulletin, U.S. Department of Health and Human Services, Public Health Service, Centers for Disease Control, National Institute for Occupational Safety and Health; December 1982.

13. *Personal Protective Equipment for Hazardous Materials Incidents: A Selection Guide*; U.S. Department of Health and Human Services, Public Health Service, Centers for Disease Control, National Institute for Occupational Safety and Health; October 1984.

14. *Fire Service Emergency Management Handbook,* International Association of Fire Chiefs Foundation, 101 East Holly Avenue, Unit 10B, Sterling, VA 22170, January 1985.

15. *Emergency Response Guidebook,* U.S. Department of Transportation, Washington, DC, 1987.

16. *Report to the Congress on Hazardous Materials Training, Planning and Prepared-*

ness, Federal Emergency Management Agency, Washington, DC, July 1986.

17. *Workbook for Fire Command,* Alan V. Brunacini and J. David Beageron, National Fire Protection Association, Batterymarch Park, Quincy, MA 02269, 1985.

18. *Fire Command,* Alan V. Brunacini, National Fire Protection, Batterymarch Park, Quincy, MA 02269, 1985.

19. *Incident Command System,* Fire Protection Publications, Oklahoma State University, Stillwater, OK 74078, 1983.

20. *Site Emergency Response Planning,* Chemical Manufacturers Association, Washington, DC 20037, 1986.

21. *Hazardous Materials Emergency Planning Guide,* NRT-1, Environmental Protection Agency, Washington, DC, March 1987.

22. *Community Teamwork: Working Together to Promote Hazardous Materials Transportation Safety,* U.S. Department of Transportation, Washington, DC, May 1983.

23. *Disaster Planning Guide for Business and Industry,* Federal Emergency Management Agency, Publication No. FEMA 141, August 1987.

(The Office of Management and Budget has approved the information collection requirements in this section under control number 1218-0139)

INDEX

D

E

F

N

O

P

R

S